State-of-the-Art Biosensors in China

State-of-the-Art Biosensors in China

Editor

Ping Yu

Basel • Beijing • Wuhan • Barcelona • Belgrade • Novi Sad • Cluj • Manchester

Editor
Ping Yu
Chinese Academy of Sciences
Beijing, China

Editorial Office
MDPI
St. Alban-Anlage 66
4052 Basel, Switzerland

This is a reprint of articles from the Special Issue published online in the open access journal *Biosensors* (ISSN 2079-6374) (available at: https://www.mdpi.com/journal/biosensors/special_issues/biosensors_cn).

For citation purposes, cite each article independently as indicated on the article page online and as indicated below:

Lastname, A.A.; Lastname, B.B. Article Title. *Journal Name* **Year**, *Volume Number*, Page Range.

ISBN 978-3-0365-8872-8 (Hbk)
ISBN 978-3-0365-8873-5 (PDF)
doi.org/10.3390/books978-3-0365-8873-5

© 2023 by the authors. Articles in this book are Open Access and distributed under the Creative Commons Attribution (CC BY) license. The book as a whole is distributed by MDPI under the terms and conditions of the Creative Commons Attribution-NonCommercial-NoDerivs (CC BY-NC-ND) license.

Contents

About the Editor vii

Preface ix

Xin Li, Xiaoling Wang, Wei Guo, Yunfei Wang, Qing Hua, Feiyan Tang, et al.
Selective Detection of Alkaline Phosphatase Activity in Environmental Water Samples by Copper Nanoclusters Doped Lanthanide Coordination Polymer Nanocomposites as the Ratiometric Fluorescent Probe
Reprinted from: *Biosensors* **2022**, *12*, 372, doi:10.3390/bios12060372 1

Chenchen Li, Jinghui Yang, Rui Xu, Huan Wang, Yong Zhang and Qin Wei
Progress and Prospects of Electrochemiluminescence Biosensors Based on Porous Nanomaterials
Reprinted from: *Biosensors* **2022**, *12*, 508, doi:10.3390/bios12070508 13

Jianping Zheng, Xiaolin Xu, Hanning Zhu, Zhipeng Pan, Xianghui Li, Fang Luo and Zhenyu Lin
Label-Free and Homogeneous Electrochemical Biosensor for Flap Endonuclease 1 Based on the Target-Triggered Difference in Electrostatic Interaction between Molecular Indicators and Electrode Surface
Reprinted from: *Biosensors* **2022**, *12*, 528, doi:10.3390/bios12070528 35

Ying Luo, Na Wu, Linyu Wang, Yonghai Song, Yan Du and Guangran Ma
Biosensor Based on Covalent Organic Framework Immobilized Acetylcholinesterase for Ratiometric Detection of Carbaryl
Reprinted from: *Biosensors* **2022**, *12*, 625, doi:10.3390/bios12080625 45

Cheng Zhou, Zecai Lin, Shaoping Huang, Bing Li and Anzhu Gao
Progress in Probe-Based Sensing Techniques for In Vivo Diagnosis
Reprinted from: *Biosensors* **2022**, *12*, 943, doi:10.3390/bios12110943 57

About the Editor

Ping Yu

Ping Yu, Ph.D., is currently a Professor of Key Laboratory of Analytical Chemistry for Living Biosystems at the Institute of Chemistry, the Chinese Academy of Sciences (ICCAS). She has worked on ion transport in confined spaces and brain neuro-electroanalytical chemistry, performing in vivo electroanalysis of brain chemicals and single-entity analysis. She received a B.Sc. in chemistry from Yantai Normal University (China) in 2001 and an M.S. in chemistry from Xiangtan University in 2004. She finished Ph.D. training in chemistry at ICCAS in 2007. She was a recipient of the "National Excellent Young Scholars" (2013) and the "National Distinguished Young Scholars" awards from the National Natural Science Foundation of China (2021), as well as the "National Distinguished Young Scholars" award from the Beijing Municipal Natural Science Foundation (2019).

Preface

Biosensors have been under development for over 54 years, and research in this field has been very popular for the last 34 years. At the beginning of this period, biosensors were known as bioelectrodes, enzyme electrodes or biocatalytic membrane electrodes. Then, the definition of biosensors broadened to include sensors buried within large automated instruments, including mass spectrometry, chromatography and electrophoresis. Nowadays, a biosensor means a device or probe that integrates a biological element, such as an enzyme or antibody, into an electronic component to generate a measurable signal. The biosensors can be classified based on bioreceptors (e.g., enzyme, immunity, aptamer, nucleic acid, microbial or whole-cell), transducers (e.g., electrochemical, electronic, thermal, optical and mass-based or gravimetric), detection systems (e.g., optical, electrical, electronic, thermal, mechanical and magnetic) and technology (surface plasmon resonance (SPR), biosensors-on-chip (lab-on-chip), micro/nanofluid and fluorescence).

We published this Special Issue entitled "State-of-the-Art Biosensors in China", which is a regional project aiming to collect high-quality research articles, comprehensive reviews and communications regarding all aspects of biosensors and biosensing in China. Our aim is to encourage scientists to publish their experimental and theoretical results in as much detail as possible, including, but not being limited to, the following topics: DNA/RNA chips; DNA/RNA sensors; Enzyme-based sensors; Lab-on-a-chip technology; micro/nanofluidic devices; immunosensors; biomaterials; real-time and in situ assay based on biosensors technology; label-free biosensors; in vivo and in vitro analysis; electroanalysis; bioelectrochemistry; nanopore sensors.

This Special Issue is composed of five papers. Firstly, Zhao et al. prepared the copper nanoclusters doped in lanthanide coordination polymer nanocomposites as the ratiometric fluorescent probe to realize the selective detection of alkaline phosphatase activity in environmental water samples. Then, Wei et al. reviewed the progress and prospects of electrochemiluminescence biosensors based on porous nanomaterials (e.g., mesoporous silica, metal-organic frameworks, covalent organic frameworks and metal-polydopamine frameworks). Next, Lin et al. developed a label-free and homogeneous electrochemical biosensor to perform flap endonuclease 1 based on the target-triggered difference in electrostatic interaction between molecular indicators and the electrode surface. Then, Ma et al. developed a ratiometric electrochemical biosensor for the detection of carbaryl based on a covalent organic framework loaded with acetylcholinesterase. Finally, Gao et al. reviewed the progress in developing probe-based sensing techniques (e.g., force, temperature, chemical and biomarker sensing) for in vivo diagnosis in China in recent years.

In conclusion, I would like to take this opportunity to express our most profound appreciation to the MDPI Book staff, the Editorial Board of the journal Biosensors, and the Assistant Editor of this Special Issue.

Ping Yu
Editor

Article

Selective Detection of Alkaline Phosphatase Activity in Environmental Water Samples by Copper Nanoclusters Doped Lanthanide Coordination Polymer Nanocomposites as the Ratiometric Fluorescent Probe

Xin Li [1], Xiaoling Wang [1], Wei Guo [2,*], Yunfei Wang [1], Qing Hua [1], Feiyan Tang [1], Feng Luan [1], Chunyuan Tian [1], Xuming Zhuang [1,*] and Lijun Zhao [1,*]

1. College of Chemistry and Chemical Engineering, Yantai University, Yantai 264005, China; lixinxin970519@163.com (X.L.); wxl18354380401@163.com (X.W.); yunfeiw0712@163.com (Y.W.); 13844630958@163.com (Q.H.); tangfeiyan0807@163.com (F.T.); luanf@ytu.edu.cn (F.L.); cytian@ytu.edu.cn (C.T.)
2. Shandong Dyne Marine Biopharmaceutical Co., Ltd., Weihai 264300, China
* Correspondence: guodawei0298@163.com (W.G.); xmzhuang@iccas.ac.cn (X.Z.); zhaoljytu@ytu.edu.cn (L.Z.)

Abstract: In this paper, a novel, accurate, sensitive and rapid ratiometric fluorescent sensor was fabricated using a copper nanoclusters@infinite coordination polymer (ICP), specifically, terbium ion-guanosine 5′-disodium (Cu NCs@Tb-GMP) nanocomposites as the ratiometric fluorescent probe, to detect alkaline phosphatase (ALP) in water. The fluorescence probe was characterized by scanning electron microscopy, transmission electron microscopy, X-ray photoelectron spectroscopy, and Fourier transform infrared spectroscopy. The experimental results showed that, compared with Tb-GMP fluorescent sensors, Cu ratiometric fluorescent sensors based on NCs encapsulated in Tb-GMP had fewer experimental errors and fewer false-positive signals and were more conducive to the sensitive and accurate detection of ALP. In addition, the developed fluorescent probe had good fluorescence intensity, selectivity, repeatability and stability. Under optimized conditions, the ratiometric fluorescent sensor detected ALP in the range of 0.002–2 U mL^{-1} (R^2 = 0.9950) with a limit of detection of 0.002 U mL^{-1}, and the recovery of ALP from water samples was less than 108.2%. These satisfying results proved that the ratiometric fluorescent probe has good application prospects and provides a new method for the detection of ALP in real water samples.

Keywords: alkaline phosphatase; copper nanoclusters; infinite coordination polymer; ratiometric; fluorescent probe

1. Introduction

In recent years, water eutrophication [1,2], which is caused by the excessive enrichment of nutrients such as nitrogen and phosphorus in water, has attracted increasing attention from society. Accumulating evidence suggests that alkaline phosphatase (ALP) not only provides available phosphorus to organisms in water but also plays a key role in the phosphorus cycle in water [3–5]. The ALP can be used as the index of phosphorus deficiency; the activity of ALP increases by 25 times when phosphorus deficiency occurs [6]. Meanwhile, ALP is an essential enzyme in phosphate metabolism because it can effectively catalyze hydrolysis or the transphosphorylation of phosphate. Therefore, accurately detecting the concentration of ALP in water, which is more important than other indicators used to evaluate water quality, is of great significance for ecological environments and human production activities [7–9]. To date, a variety of methods for detecting ALP, such as colorimetric [10], electrochemical [11], surface-enhanced Raman dispersion [12], and molecular fluorescence methods [13], have been reported. Among these methods, molecular fluorescence has attracted extensive attention due to its advantages of fast detection speed, small sample quantity and high sensitivity.

In the molecular fluorescence method, ratiometric fluorescent sensors have been widely used for the past few years because of their high selectivity and accuracy [14,15]. This method mainly analyzes the change in the ratios of the signal intensities at different fluorescence wavelengths, which can effectively reduce the error and false-positive signals in the experiment [16]. In recent years, the construction of ratiometric fluorescent sensors has attracted the attention of many researchers. Loas and his colleagues developed a Cu-based ratiometric fluorescent sensor for nitric oxide detection that operated on the energy transfer between hydroxycoumarin and luciferin chromophores [17]. Wang and coworkers designed a dual-emission ratiometric fluorescent sensor to detect folic acid by doping ZnS quantum dots with Cu^{2+} and Mn^{2+} ions [18]. Ye and his workmates developed a ratiometric fluorescent platform for the amplification of kanamycin without an enzyme signal to enable the detection of antibiotics [19]. Although many nanocomposites have been used in ratiometric fluorescent sensors, the development of a novel kind of nanocomposite with excellent performance is still worth studying.

To date, an infinite coordination polymer (ICP) nanocomposite based on terbium ions (Tb^{3+}) and guanosine 5′-disodium (GMP) has been developed [20]. Compared with other coordination polymers, ICPs have obvious advantages, such as high structural flexibility [21], a good guest envelope, and a fast response to external stimuli [22–24]. Meanwhile, copper nanoclusters (Cu NCs) have gradually become a research and application hotspot in the fields of biological analysis and environmental monitoring with broad prospects because of their good biocompatibility, low biotoxicity, good photostability and low cost [25–27]. However, Cu NCs also have disadvantages such as easy oxidation and aggregation in the preparation process [25]. To overcome these difficulties, researchers have proposed many different solutions. Some researchers have used different materials to synthesize Cu NCs [28–30], whereas others have used ZIFs and other materials to encapsulate Cu NCs to form nanocomposites [31–33].

In this paper, a ratiometric fluorescent probe named Cu NCs@Tb-GMP was successfully prepared and could detect ALP quickly. On the one hand, Cu NCs could be used to sensitize Tb-GMP to enhance its fluorescence intensity; on the other hand, using Cu NCs as an internal standard [34] could effectively correct the error and improve the accuracy of the experiment. The mechanism is shown in Scheme 1. In the presence of ALP, the phosphate group in GMP could be hydrolyzed by ALP, leading to the destruction of the Tb-GMP network, and the characteristic emission intensity of Tb-GMP at 545 nm was significantly reduced. At the same time, due to the hydrolysis of the polymer network, Cu NCs were released into the solution, resulting in a slight but almost negligible increase in their fluorescence at 425 nm [34,35]. Thus, the ratio of fluorescent intensity at 545 nm and 425 nm has been choosing as the ratiometric fluorescent parameter value. After a series of characterizations and condition optimization, the Cu NCs@Tb-GMP probe was successfully applied to the detection of ALP in environmental water samples under the optimal conditions. According to the experimental results, compared with other methods, the constructed molecular fluorescent probe had the advantages of higher stability, better sensitivity and excellent anti-interference ability. At the same time, the preparation method of the synthesized fluorescent probe was simple and the cost was relatively low; therefore, it has wide application prospects in the protection of ecological environmental systems and human health.

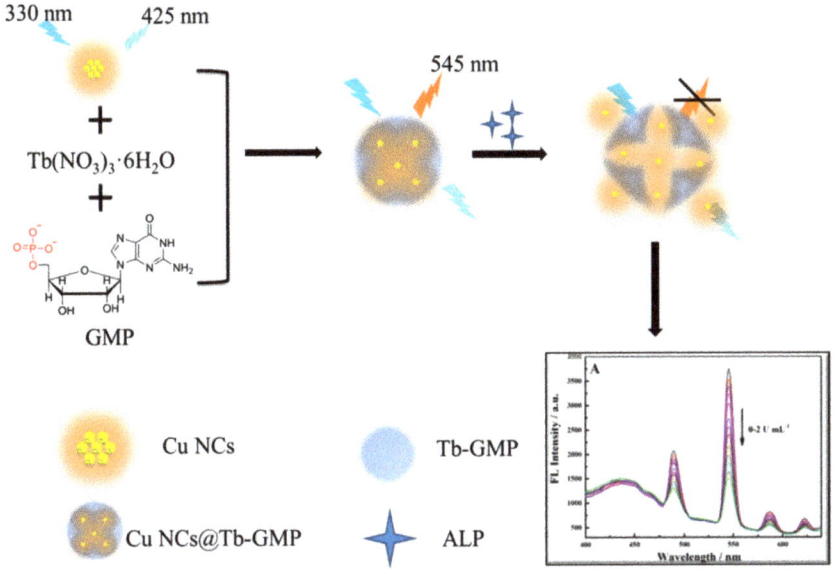

Scheme 1. A schematic of the mechanism of Cu NCs@Tb-GMP for ALP detection.

2. Materials and Methods

2.1. Materials

GMP, terbium nitrate hexahydrate (Tb(NO$_3$)$_3 \cdot$ 6H$_2$O), n-2-hydroxyethylpiperazine-n'-2-ethanesulfonic acid (HEPES), ALP, glucose dehydrogenase (GDH), thrombin, glucose oxidase (GOx) and horseradish peroxidase (HRP) were purchased from Shanghai Aladdin Biochemical Technology Co., Ltd. (Shanghai, China). Copper sulfate pentahydrate (CuSO$_4$ $\cdot$ 5H$_2$O), ethylenediamine (EDA) and L-ascorbic acid (AA) were obtained from Sinopharm Chemical Reagent Co., Ltd. (Tianjin, China). Phosphate buffer solution (PBS, 0.1 M, pH = 7.4) was prepared from a standard mixture of sodium dihydrogen phosphate and dipotassium hydrogen phosphate. All the solutions involved in these experiments were prepared with ultrapure water (18.25 MΩ cm).

2.2. Apparatus

Scanning electron microscopy (SEM) images and transmission electron microscopy (TEM) images were obtained on JSM-7900F and JEM-2100 (200 kV) instruments, respectively (JEOL Ltd., Tokyo, Japan). High-resolution transmission electron microscopy (HR-TEM) images were obtained on a JEM-2100F transmission electron microscope at an accelerating voltage of 200 kV (JEOL Ltd., Tokyo, Japan). X-ray photoelectron spectroscopy (XPS) measurements were performed on a Thermo ESCALAB-250 instrument (Thermo Fisher Scientific, Waltham, MA, USA). The fluorescence emission spectra were obtained on an F-4700 fluorescence spectrophotometer (HITACHI, Tokyo, Japan). A Nicolet 5700 Fourier transform infrared (FT-IR) spectrometer (Thermo Electron Corporation, Waltham, MA, USA) was used to obtain FT-IR spectra. Ultraviolet-visible absorption spectroscopy (UV-vis) was performed with an A560 ultraviolet-visible spectrometer (AOE Instruments, Shanghai, China) at wavelength intervals of 2 nm.

2.3. Synthesis of Nanocomposites

2.3.1. Synthesis of Cu NCs

Cu NCs were prepared according to a simple approach reported previously [25]. The specific steps were as follows: first, CuSO$_4 \cdot$ 5H$_2$O (0.08 mM), EDA (1.36 mM) and AA

(0.800 mM) were dissolved in 24 mL of ultrapure water and adjusted with 1 M NaOH to keep the pH at approximately 4.5. Next, the solution obtained above was heated to 37 °C and reacted at a constant temperature for 30 min. Then, the solution was cooled to room temperature, transferred to a centrifuge tube and centrifuged at 4000 rpm for 10 min. After washing 3 times, the supernatant was removed and the precipitate was dispersed in ultrapure water to obtain the required Cu NCs, which were stored at 4 °C for further experiments.

2.3.2. Synthesis of Tb-GMP and the Cu NCs@Tb-GMP Ratiometric Fluorescent Probe

According to reference [20], Tb-GMP and Cu NCs@Tb-GMP were obtained. The concentrations of $Tb(NO_3)_3 \cdot 6H_2O$ and the GMP solution were both 10 mM. When the two solutions were mixed in equal volumes at room temperature, a white precipitate was obtained, washed with ultrapure water and centrifuged at 5000 rpm for 10 min. In this way, Tb-GMP was successfully synthesized. Cu NCs@Tb-GMP was obtained by a similar method. For the synthesis of the Cu NCs@Tb-GMP ratiometric fluorescent probe, the above synthesis steps were changed slightly to ensure encapsulation of Cu NCs in the Tb-GMP. The only difference was that Cu NCs were added to the $Tb(NO_3)_3 \cdot 6H_2O$ solution during the Tb-GMP synthesis process to fully encapsulate Cu NCs in Tb-GMP.

2.4. Construction of the Ratiometric Fluorescent Sensor

A series of experiments was designed to develop ratiometric fluorescent sensors. First, an aqueous solution of ALP with an activity of 10 U mL^{-1} was prepared. In addition, a certain volume of the ALP aqueous solution was successively added to the Cu NCs@Tb-GMP ratiometric fluorescent probe, which resulted in an ALP activity in the Cu NCs@Tb-GMP aqueous solution of 0–2 U mL^{-1}. Then, the mixed solution was placed in a thermostatic water bath at 37 °C for 30 min, ensuring full reaction. A fluorescence spectrophotometer was selected as the detection method. In this way, a ratiometric fluorescent sensor that could achieve sensitive detection of ALP in aqueous solutions was successfully fabricated.

2.5. Fluorescence Assay of ALP in Real Samples

Different samples were obtained from local river in Yantai City (water samples were taken from the same river in upper, middle and lower reach area) to evaluate the performance of the proposed method. First, the sample through a 0.45 μm mem-143 brane was filtered for further use. Next, the sample was diluted 10 times with PBS (0.1 M, pH = 7.4), which was stored at −20 °C for further use.

3. Results and Discussion

3.1. Characterization of the Cu NCs, Tb-GMP and Cu NCs@Tb-GMP

The microstructures, elements and optical properties of the Cu NCs, Tb-GMP and Cu NCs@Tb-GMP were investigated. The morphologies of the Cu NCs, Tb-GMP and Cu NCs@Tb-GMP were characterized by TEM and SEM. Figure 1A and the inset showed that the Cu NCs exhibited a random distribution characteristic of excellent dispersion, and the diameter distribution of the Cu NCs was approximately 4.67 ± 1.12 nm (Figure 1B). SEM images of the ICP and Cu NCs@Tb-GMP are shown in Figure 1C,D, from which some information about the microstructures could be obtained. Tb-GMP was clearly shown to be a spherical colloid. No obvious changes were observed in the microstructure when the Cu NCs were added to Tb-GMP, which proved that the addition of the Cu NCs had no effect on the morphology of Tb-GMP. Thus, a ratiometric fluorescent probe with an ICP network based on Tb and GMP containing Cu NCs was successfully prepared, and the Cu NCs could be used to sensitize Tb-GMP, enhancing its fluorescence intensity. Furthermore, using Cu NCs as an internal standard [36] could effectively correct the error and improve the accuracy of the experiments.

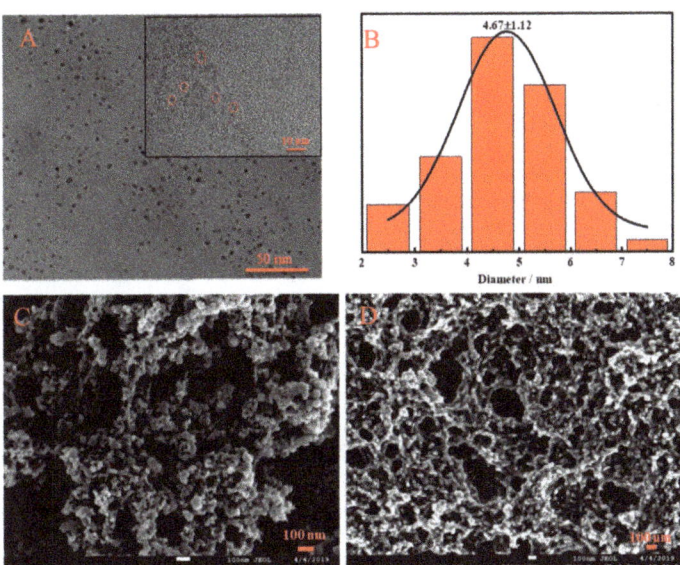

Figure 1. (**A**) A TEM image of the Cu NCs (inset shows an HR-TEM image of the Cu NCs) and (**B**) the diameter distribution of the Cu NCs. (**C**,**D**) are SEM images of Tb-GMP and Cu NCs@Tb-GMP, respectively.

XPS was used to explore the elemental compositions of the synthetic materials, as shown in Figure 2A. All the expected elements were present, including C, N, O, Cu and Tb, and their characteristic peaks were located at 285.51, 398.93, 530.97, 950.87 and 1073.23 eV, respectively. Three peaks could be identified in the high-resolution XPS spectrum of Cu (inset). The peaks located at 932.16, 940.83 and 951.89 eV could be assigned to Cu $2p_{2/3}$ and Cu $2p_{1/2}$ [30]. The XPS results showed that the Cu NCs@Tb-GMP nanocomposite contained all the elements of the raw materials, which proved that it had been successfully synthesized [16]. The fluorescence properties of the Cu NCs (red curve), Tb-GMP (black curve) and Cu NCs@Tb-GMP (blue curve) at an excitation wavelength of 330 nm are shown in Figure 2B. The red curve shows that the maximum emission wavelength of the Cu NCs was approximately 425 nm under the optimal excitation wavelength of 330 nm. The black curve provided some information indicating that the emission wavelengths of Tb-GMP situated at 489, 545, 587 and 642 nm due to the coordination of O6 and N7 moieties to Tb^{3+} promoted the energy transfer from the guanine base to the emissive D_4 state of the Tb^{3+} in the process of self-assembly. To verify whether the Cu NCs were coated with Tb-GMP, the blue curve that we obtained was compared with the red and black curves, and no overlap between the emission wavelengths of the Cu NCs and Tb-GMP was observed at the same excitation wavelength, demonstrating that the Cu NCs were successfully encapsulated in the Tb-GMP network.

Figure 2C shows the FT-IR spectra of the Cu NCs (black curve), Tb-GMP (blue curve) and Cu NCs@Tb-GMP (red curve). The characteristic peaks of the Cu NCs at 3550 and 3300 cm^{-1} were attributed to amino groups (NH_2). However, the peaks were not obvious, indicating that EDA was connected with the Cu core through NH_2, which was consistent with a previous reference (black curve) [25]. As shown in the blue curve, the strong absorption at 1537 cm^{-1} was attributed to the pyrimidine ring vibration of GMP. The absorption at 603 cm^{-1} was attributed to the nonsymmetrical bending of PO_4^{3+} in GMP. The weak absorption peak at 1388 cm^{-1} came from the pyrimidine ring of GMP. The peak at 1164 cm^{-1} may have been the result of the vibration characteristics of the GMP sugar ring. FT-IR spectroscopy proved that Tb/GMP was synthesized successfully. In the

FT-IR spectrum of Cu NCs@Tb-GMP (red curve), we observed the previously mentioned characteristic peaks. Based on the previous analysis, the Cu NCs@Tb-GMP nanocomposites were successfully synthesized [37]. Figure 2D shows the UV-vis spectra of the Cu NCs (black curve), Tb-GMP (red curve) and Cu NCs@Tb-GMP (blue curve). The black curve had two peaks at 370 nm and 560 nm. The absorption peak at 370 nm was attributed to the presence of NH_2 in the Cu NCs. The characteristic peak at 560 nm was consistent with the characteristic surface plasmon resonance band of the Cu NCs. The absorption peaks of the red and black curves at 240 nm were caused by the π-π* transition in the nanocomposite, whereas the disappearance of the absorption peak of the black curve indicated that the Cu NCs were successfully encapsulated in Tb-GMP, which further proved the successful synthesis of the composite materials and was consistent with previous research [38].

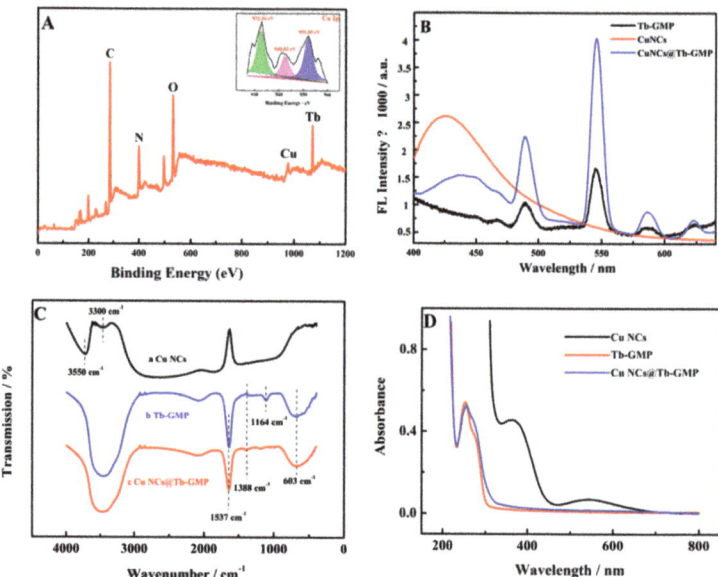

Figure 2. (**A**) The XPS spectrum of Cu NCs@Tb-GMP with the Cu 2p region (inset). (**B**) The fluorescence properties of the Cu NCs (red curve), Tb-GMP (black curve) and Cu NCs@Tb-GMP (blue curve). (**C**) The FT-IR spectra of the Cu NCs (black curve), Tb-GMP (blue curve) and Cu NCs@Tb-GMP (red curve). (**D**) The UV–vis absorption spectra of the Cu NCs (black curve), Tb-GMP (red curve) and Cu NCs@Tb-GMP (blue curve).

3.2. Optimization Assay

To explore the volume ratio of Cu NCs to Tb-GMP in the process of synthesis, a series of experiments were designed. The amount of Tb-GMP remained the same, while the amount of Cu NCs was changed to obtain volume ratios of 10:3, 10:4, 10:5, 10:6 and 10:7. The results are shown in Figure 3A, from which a conclusion could be obtained. When the volume ratio of Tb-GMP to Cu NCs was 10:5, the fluorescence intensity of the Cu NCs@Tb-GMP nanocomposite was the strongest, and the emission peaks of the Cu NCs and Tb-GMP existed simultaneously. Based on the above experiments, 10:5 was chosen as the optimal volume ratio for the synthesis of Cu NCs@Tb-GMP in the subsequent experiments; when the content of Cu NCs was low, they were insufficient to sensitize Tb-GMP, whereas when the content of Cu NCs was too high, the fluorescence characteristics of Tb-GMP were good, with an inconspicuous Cu NCs emission peak. After the synthesis of Tb-GMP, experiments were conducted to determine the optimal excitation wavelength. Excitation wavelengths of

330, 340, 350 and 360 nm were employed in the experiments (Figure 3B), and 330 nm was found to be the optimal excitation wavelength of Tb-GMP.

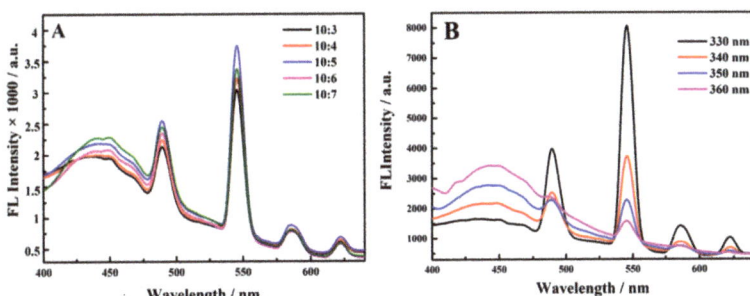

Figure 3. (**A**) The Cu NCs were added to Tb-GMP in different ratios. (**B**) The determination of the optimal excitation wavelength of Cu NCs@Tb-GMP.

3.3. Linearity of the ALP Ratiometric Fluorescent Sensor in ALP Detection

Under the optimal experimental conditions, different concentrations of ALP were added to the prepared Cu NCs@Tb-GMP, and the fluorescence response was detected after a 30 min reaction at 37 °C. The mechanism of the ratiometric fluorescent sensor is shown in Scheme 1. For the Cu NCs@Tb-GMP nanocomposites under an excitation wavelength of 330 nm, the emission wavelengths of the nanocomposite were 425 and 545 nm. For Cu NCs@Tb-GMP in the presence of ALP, the phosphate group in GMP could be hydrolyzed, leading to the structural damage of the Cu NCs@Tb-GMP, leading to the wavelength at 545 nm decreasing. Simultaneously, the Cu NCs were released from the Cu NCs@Tb-GMP, while the wavelength at 425 nm had no effect, which may have been due to the surface of Tb-GMP possessing so many Cu NCs that the Cu NCs after being released from Cu NCs@Tb-GMP uninfluenced the fluorescence intensity. The ratio of the two peaks was used to construct a relationship. The results are shown in Figure 4A, from which much information was obtained. The fluorescence intensity of Tb-GMP decreased (545 nm) as the concentration of ALP increased between 0–2 U mL^{-1}, whereas the fluorescence intensity of the Cu NCs remained unchanged (425 nm). The inset in Figure 3A displays the corresponding images of ALP solutions with different concentrations under 365 nm UV light (original from the UV flashlight the fixed wavelength). The color of Cu NCs@Tb-GMP became lighter (from left to right), proving that the degree of quenching became stronger with increasing ALP concentration. Figure 4B shows the linear relationship between F_{545}/F_{425} and the ALP concentration. The linear equation was $F_{545}/F_{425} = 2.694 - 0.8214\, C_{ALP}$ and $R^2 = 0.9950$, and the detection limit (LOD) was 0.002 U mL^{-1}, which suggested that the prepared sensor based on the Cu NCs@Tb-GMP ratiometric fluorescent probe could detect ALP sensitively and accurately. As shown in Figure 4C, the Tb-GMP detects the various concentrations of ALP, and the fluorescence intensity of Tb-GMP was nearly lower than three times that of Cu NCs@Tb-GMP at 545 nm, demonstrating that the Cu NCs possess the ability to sensitize Tb-GMP to improve the intensity. Compared with other types of ALP probes (Table 1), the prepared Cu NCs@Tb-GMP had a lower detection limit and a wider detection range, attesting that the prepared sensor based on the Cu NCs@Tb-GMP ratiometric fluorescent probe had better practicability and could be more effective at detecting ALP in real samples.

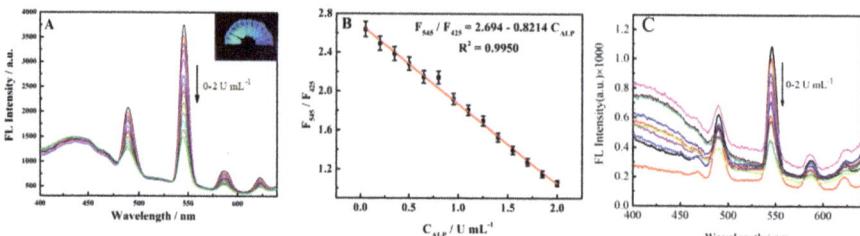

Figure 4. (**A**) The response of the Cu NC@Tb-GMP sensor when exposed to ALP solutions with different concentrations. The inset displays the corresponding images of ALP solutions with different concentrations under 365 nm UV light. (**B**) The plot of the response of the Cu NC@Tb-GMP ratiometric fluorescent sensor as a function of the concentration of ALP. (**C**) The response of the Tb-GMP sensor when exposed to ALP solutions with different concentrations.

Table 1. A comparison of ALP detection of different kind of fluorescent probes.

Nanoprobes	Linear Range/U mL^{-1}	Detection Limit/U mL^{-1}	Ref.
coumarin@Tb-GMP [a]	0.025–0.2	0.01	[39]
AuNPs/GO [b]	0.1–1	0.009	[40]
ATP-Cu [c]	0.03–0.3	0.03	[41]
Cu(BCDS [d])$_2^{2-}$	0.027–0.220	0.027	[42]
Cu NCs@Tb-GMP	0.002–2	0.002	This work

[a] Terbium-guanine monophosphate; [b] Gold nanoparticles/graphene oxide; [c] Adenosine triphosphate -copper nanozymes; [d] Bathocuproine disulfonate.

3.4. Stability, Salt Tolerance, Selectivity and Repeatability of the Probe

Cu NCs possess outstanding stability due to this synthesis method [25]. The Cu NCs could be stable in Tb-GMP. A portion of Cu NCs has been encapsulated into Tb-GMP, and others are located on the surface of Tb-GMP, due to the Cu NCs@Tb-GMP synthesis process. The Cu NCs connect with Tb-GMP were not through any chemical bonds. Therefore, Cu NCs could sustain the stability in the Cu NCs@Tb-GMP. To explore the stability of Cu NCs@Tb-GMP, the synthesized nanocomposite was stored at 4 °C for 7 days, and its fluorescence intensity was detected and recorded at the same excitation wavelength. As shown in Figure 5A, the fluorescence intensity of Cu NCs@Tb-GMP decreased slightly after 7 days but remained basically unchanged. The inset shows the stability of Cu NCs@Tb-GMP over seven days, indicating that the ratiometric fluorescent probe synthesized in the experiment had high stability and would not undergo structural changes for a long time at 4 °C. Meanwhile, experiments were performed to further investigate the salt tolerance of Cu NCs@Tb-GMP, that is, a certain amount of NaCl was added to the synthesized ratiometric fluorescent probe to gradually increase its concentration from 0 to 100 mM, and at the same time, the fluorescence intensity was measured and recorded under a fixed excitation wavelength. As shown in Figure 5B, the fluorescence intensity of Cu NCs@Tb-GMP did not change significantly with increasing NaCl concentration, which indicated that the synthesized Cu NCs@Tb-GMP ratiometric fluorescent probe still had excellent stability in a high concentration salt solution.

In addition, the selectivity of Cu NCs@Tb-GMP for ALP detection was further studied. Other enzymes that may exist in freshwater lakes, such as GDH, thrombin, GOx and HRP, were selected as interference substances to conduct control experiments under the same experimental conditions. The results are shown in Figure 5C. The effect of the interference substance alone on Cu NCs@Tb-GMP was almost negligible, but when the interference substance and ALP were added at the same time, an obvious quenching phenomenon occurred. The results indicated that only the ALP could enable the hydrolysis of the phosphate radical, leading to the structure and energy transmission of Cu NCs@Tb-GMP being damaged. Additionally, the prepared Cu NCs@Tb-GMP had good anti-interference ability during the detection of ALP and could be used for detection in real water samples. Based on this, we selected Cu NCs@Tb-GMP toward ALP over other chosen interference substances.

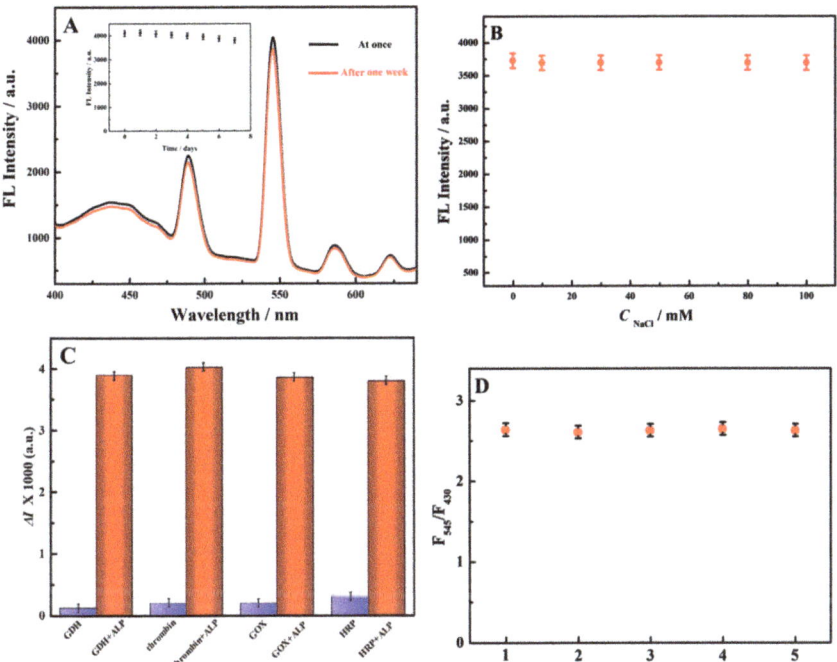

Figure 5. (**A**) The stability of Cu NCs@Tb-GMP after one week; the inset shows the stability each day of the week. (**B**) The salt tolerance of Cu NCs@Tb-GMP. (**C**) The selectivity of other interfering substances over ALP: GDH, thrombin, GOx and HPR. (**D**) The repeatability of Cu NCs@Tb-GMP over five experiments.

Moreover, some experiments were designed to verify the reproducibility of Cu NCs@Tb-GMP under the same conditions. The satisfactory results are shown in Figure 5D. In many experiments, the ratio of the fluorescence intensities of the synthesized Cu NCs@Tb-GMP ratiometric fluorescent probe at 545 nm and 425 nm did not vary considerably, which proved that the synthesized probe had good repeatability.

3.5. Detection of ALP in Real Samples

Water quality ensures people's health, and the content of ALP in water is more important than other indicators used to evaluate water quality. Therefore, the accurate and sensitive determination of ALP in water is vital for evaluating water quality. Three samples, which were obtained from a local river (Yantai City), were prepared, and the results are displayed in Table 2. The recovery was between 96.7% and 108.2%, and the relative standard deviation of recovery was obtained by each water sample have three parallel experiments was less than 3%. The experimental results were satisfactory, and it is expected that Cu NCs@Tb-GMP can be applied to ALP detection in various water samples from the environment.

Table 2. The recoveries for ALP determination in samples.

Sample	Add/U mL^{-1}	Founded/U mL^{-1}	Recovery/%
1	0.2000	0.2164	108.2
	0.5000	0.5036	100.7
2	0.2000	0.2120	106.0
	0.5000	0.4982	99.64
3	0.2000	0.2127	106.4
	0.5000	0.4835	96.70

4. Conclusions

In this paper, Cu NCs were encapsulated in an ICP based on Tb^{3+} and GMP nanocomposites through a certain reaction to form a new type of ratiometric fluorescent probe called Cu NCs@Tb-GMP. A ratiometric fluorescent sensor strategy for the detection of ALP in water samples was developed based on this ratiometric fluorescent probe. Compared with simple Tb-GMP, Cu NCs@Tb-GMP had much higher fluorescence intensity, and its salt resistance and stability were satisfactory. When ALP was added to Cu NCs@Tb-GMP, the fluorescence intensity corresponding to Tb-GMP showed a good linear relationship with the change in ALP, but the fluorescence corresponding to the Cu NCs basically remained unchanged. This satisfying result proved that the ratiometric fluorescent sensor constructed in this paper had good stability, good reproducibility and high selectivity for ALP and provides a new way to detect water eutrophication.

Author Contributions: Conceptualization, X.L., X.W. and Y.W.; methodology, X.L., X.W., C.T. and X.Z.; software, X.L., X.W., F.L., C.T. and X.Z. validation, X.L., X.W., F.L., C.T. and X.Z.; formal analysis, X.L., W.G., Q.H. and F.T.; investigation, C.T., W.G., L.Z. and X.Z.; resources, W.G., C.T., L.Z. and X.Z.; writing—original draft preparation, X.L, L.Z. and X.W.; and writing—review and editing, W.G., F.L., C.T. and X.Z.; All authors have read and agreed to the published version of the manuscript.

Funding: This research was funded by the National Natural Science Foundation of China, grant number 21778047.

Institutional Review Board Statement: Not applicable.

Informed Consent Statement: Not applicable.

Data Availability Statement: Not applicable.

Conflicts of Interest: There are no conflict to declare.

References

1. Budria, A. Beyond troubled waters: The influence of eutrophication on host–parasite interactions. *Funct. Ecol.* **2017**, *31*, 1348–1358. [CrossRef]
2. Pakhomova, S.; Yakushev, E.; Protsenko, E.; Rigaud, S.; Cossa, D.; Knoery, J.; Couture, R.M.; Radakovitch, O.; Yakubov, S.; Krzeminska, D. Modeling the influence of eutrophication and redox conditions on Mercury cycling at the sediment-water inter cling at the sediment-water interface in the Berr face in the Berre Lagoon. *Front. Mar. Sci.* **2018**, *5*, 291. [CrossRef]
3. Wang, J.L.; Fu, Z.S.; Qiao, H.X.; Liu, F.X. Assessment of eutrophication and water quality in the estuarine area of Lake Wuli, Lake Taihu, China. *Sci. Total. Environ.* **2018**, *650*, 1392. [CrossRef]
4. Chang, M.Q.; Teurlincx, S.; Angelis, D.L.D.; Janse, J.H.; Troost, T.A.; Wijk, D.; Mooij, W.M.; Janssen, A.B.G. A generically parameterized model of Lake eutrophication (GPLake) that links field-, lab- and model-based knowledge. *ACS Appl. Mater. Interf.* **2019**, *695*, 133887. [CrossRef]
5. Taipale, S.J.; Vuorio, K.; Aalto, S.L.; Peltomaa, E.; Tiirola, M. Eutrophication reduces the nutritional value of phytoplankton in boreal lakes. *Environ. Res.* **2019**, *179*, 108836. [CrossRef]
6. Fitzgerald, G.P.; Nelson, T.C. Extractive and enzymatic analyses for limiting or surplus phosphorus in alage. *J. Phycol.* **1966**, *2*, 32–37. [CrossRef]
7. Holtz, K.M.; Kantrowitz, E.R. The mechanism of the alkaline phosphatase reaction: Insights from NMR, crystallography and site-specific mutagenesis. *FEBS Lett.* **1999**, *462*, 7–11. [CrossRef]
8. Yang, K.C.; Metcalf, W.W. A new activity for an old enzyme: Escherichia coli bacterial alkaline phosphatase is a phosphite-dependent hydrogenase. *Proc. Natl. Acad. Sci. USA* **2004**, *101*, 7919–7924. [CrossRef]

9. Orimo, H. The Mechanism of Mineralization and the Role of Alkaline Phosphatase in Health and Disease. *J. Nippon Med. Sch.* **2010**, *77*, 4–12. [CrossRef]
10. Choi, Y.; Ho, N.; Tung, C. Sensing Phosphatase Activity by Using Gold Nanoparticles. *Angew. Chem. Int. Ed.* **2007**, *46*, 707–709. [CrossRef]
11. Kazakevičienė, B.; Valinčius, G.; Kažemėkaitė, M.; Razumas, V. Self-Assembled Redox System for Bioelectrocatalytic Assay of l-Ascorbylphosphate and Alkaline Phosphatase Activity. *Electroanalysis* **2010**, *20*, 2235–2240. [CrossRef]
12. Ingram, A.; Moore, B.D.; Graham, D. Simultaneous detection of alkaline phosphatase and beta-galactosidase activity using SERRS. *Bioorg. Med. Chem. Lett.* **2009**, *19*, 1569–1571. [CrossRef]
13. Liu, Y.; Schanze, K.S. Conjugated Polyelectrolyte-Based Real-Time Fluorescence Assay for Alkaline Phosphatase with Pyrophosphate as Substrate. *Anal. Chem.* **2008**, *80*, 8605–8612. [CrossRef]
14. Lv, Y.Y.; Cao, M.D.; Li, J.K.; Wang, J.B. A Sensitive Ratiometric Fluorescent Sensor for Zinc(II) with High Selectivity. *Sensors* **2013**, *13*, 3131. [CrossRef]
15. Liang, J.; Liu, H.B.; Wang, J. Pyrene-based ratiometric and fluorescent sensor for selective Al^{3+} detection. *Inorg. Chim. Acta* **2019**, *489*, 61–66. [CrossRef]
16. Ma, R.X.; Xu, M.; Liu, C.; Shi, G.Y.; Deng, J.J.; Zhou, T.S. Stimulus Response of GQD-Sensitized Tb/GMP ICP Nanoparticles with Dual-Responsive Ratiometric Fluorescence: Toward Point-ofUse Analysis of Acetylcholinesterase and Organophosphorus Pesticide Poisoning with Acetylcholinesterase as a Biomarker. *ACS. Appl. Mater. Inter.* **2020**, *12*, 42119. [CrossRef]
17. Loas, A.; Lippard, S.J. Direct ratiometric detection of nitric oxide with Cu(II)-based fluorescent probes. *J. Mater. Chem. B* **2017**, *5*, 8929–8933. [CrossRef]
18. Wang, Y.B.; Yang, M.; Ren, Y.K.; Fan, J. Cu-Mn codoped ZnS quantum dots-based ratiometric fluorescent sensor for folic acid. *Anal. Chim. Acta* **2018**, *1040*, 136–142. [CrossRef]
19. Ye, T.; Lu, J.Q.; Yuan, M.; Cao, H.; Yin, F.Q.; Wu, X.X.; Hao, L.L.; Xu, F. Toehold-mediated enzyme-free cascade signal amplification for ratiometric fluorescent detection of kanamycin. *Sens. Actuat. B Chem.* **2021**, *340*, 129939. [CrossRef]
20. Zhang, X.L.; Deng, J.J.; Xue, Y.M.; Shi, G.Y.; Zhou, T.S. Stimulus Response of Au-NPs@GMP-Tb Core−Shell Nanoparticles:Toward Colorimetric and Fluorescent Dual-Mode Sensing of Alkaline Phosphatase Activity in Algal Blooms of a Freshwater Lake. *Environ. Sci. Technol.* **2016**, *50*, 847–855. [CrossRef]
21. Lu, W.; Cao, X.; Tao, L.; Ge, J.; Dong, J.; Qian, W. A novel label-free amperometric immunosensor for carcinoembryonic antigen based on Ag nanoparticle decorated infinite coordination polymer fibres. *Biosens. Bioelectron.* **2014**, *57*, 219–225. [CrossRef]
22. Deng, J.; Fei, W.; Ping, Y.; Mao, L. On-site sensors based on infinite coordination polymer nanoparticles: Recent progress and future challenge. *Appl. Mater. Today* **2018**, *11*, 338–351. [CrossRef]
23. Zhang, B.; Liu, B.; Chen, G.; Tang, D. Redox and catalysis 'all-in-one' infinite coordination polymer for electrochemical immunosensor of tumor markers. *Biosens. Bioelectron.* **2015**, *64*, 6–12. [CrossRef]
24. Tan, H.; Yang, C. Ag^+-enhanced fluorescence of lanthanide/nucleotide coordination polymers and Ag^+ sensing. *Chem. Commun.* **2011**, *47*, 12373. [CrossRef]
25. Jiao, M.X.; Li, Y.; Jia, Y.X.; Xu, L.; Xu, G.Y.; Guo, Y.S.; Luo, X.L. Ligand-modulated aqueous synthesis of color-tunable copper nanoclusters for the photoluminescent assay of Hg(II). *Microchim. Acta* **2020**, *187*, 545. [CrossRef]
26. Wang, D.W.; Wang, Z.Q.; Wang, X.B.; Zhuang, X.M.; Tian, C.Y.; Luan, F.; Fu, X.L. Functionalized copper nanoclusters-based fluorescent probe with aggregation-induced emission property for selective detection of sulfide ions in food additives. *J. Agric. Food. Chem.* **2020**, *68*, 11301. [CrossRef]
27. Liao, H.; Zhou, Y.; Chai, Y.; Yuan, R. An Ultrasensitive Electrochemiluminescence Biosensor for Detection of MicroRNA by in-situ Electrochemically Generated Copper Nanoclusters as Luminophore and TiO_2 as Coreaction Accelerator. *Biosens. Bioelectron.* **2018**, *114*, 10–14. [CrossRef]
28. Yao, Z.; Liu, H.; Liu, Y.; Diao, Y.; Li, Z. FRET-based fluorometry assay for curcumin detecting using PVP-templated Cu NCs. *Talanta* **2021**, *223*, 121741. [CrossRef]
29. Geng, F.H.; Zou, C.P.; Liu, J.H.; Zhang, Q.C.; Guo, X.Y.; Fan, Y.C.; Yu, H.D.; Yang, S.; Liu, Z.P.; Li, L. evelopment of luminescent nanoswitch for sensing of alkaline phosphatase in human serum based onAl3þ-PPi interaction and CuNCs with AIE properties. *Anal. Chim. Acta* **2019**, *1076*, 131–137. [CrossRef]
30. Zhang, J.; Zhang, Z.; Ou, Y.; Zhang, F.; Meng, J.; Wang, G.; Fang, Z.L.; Li, Y. Red-Emitting GSH-Cu NCs as Triplet Induced Quenched Fluorescent Probe for Fast Detection of Thiol Pollutants. *Nanoscale* **2020**, *12*, 19429–19437. [CrossRef]
31. Cadiau, A.; Auguste, S.; Taulelle, F.; Martineau, C.; Adil, K. Hydrothermal synthesis, ab-initio structure determination and NMR study of the first mixed Cu–Al fluorinated MOF. *CrystEngComm* **2013**, *15*, 3430–3435. [CrossRef]
32. Luz, I.; Loiudice, A.; Sun, D.T.; Queen, W.L.; Buonsanti, R. Understanding the Formation Mechanism of Metal Nanocrystal@MOF-74 Hybrids. *Chem. Mater.* **2016**, *28*, 3839–3849. [CrossRef]
33. Wang, L.; Li, S.; Chen, Y.; Jiang, H.L. Encapsulating Copper Nanocrystals into Metal–Organic Frameworks for Cascade Reactions by Photothermal Catalysis. *Small* **2021**, *17*, 2004481. [CrossRef]
34. Sun, W.D.; Han, X.; Qu, F.L.; Kong, R.M.; Zhao, Z.L. A carbon dot doped lanthanide coordination polymer nanocomposite as the ratiometric fluorescent probe for the sensitive detection of alkaline phosphatase activity. *Analyst* **2021**, *146*, 2862–2870. [CrossRef] [PubMed]

35. Yang, J.O.; Li, C.Y.; Li, Y.F.; Yang, B.; Li, S. An infinite coordination polymer nanoparticles-based near-infrared fluorescent probe with high photostability for endogenous alkaline phosphatase in vivo. *J. Sensor. Actuat. B Chem.* **2018**, *225*, 3355–3363. [CrossRef]
36. Kalambet, Y.; Kozmin, Y. Internal standard arithmetic implemented as relative concentration/relative calibration. *J. Chemometr.* **2019**, *33*, e3106. [CrossRef]
37. Liu, N.; Hao, J.; Cai, K.Y.; Zeng, M.; Song, Y. Ratiometric fluorescence detection of superoxide anion based on AuNPs-BSA@Tb/GMP nanoscale coordination polymers. *Luminescence* **2017**, *33*, 119–124. [CrossRef]
38. Mott, D.; Galkowski, J.; Wang, L.; Luo, J.; Zhong, C.J. Synthesis of Size-Controlled and Shaped Copper Nanoparticles. *Langmuir* **2007**, *23*, 5740–5745. [CrossRef]
39. Deng, J.J.; Yu, P.; Wang, Y.X.; Mao, L.Q. Real-time Ratiometric Fluorescent Assay for Alkaline Phosphatase Activity with Stimulus Responsive Infinite Coordination Polymer Nanoparticles. *Anal. Chem.* **2015**, *87*, 3080–3086. [CrossRef]
40. Mahato, K.; Purohit, B.; Kumar, A.; Chandra, P. Clinically comparable impedimetric immunosensor for serum alkaline phosphatase detection based on electrochemically engineered Au-nano-dendroids and graphene oxide nanocomposite. *Biosens. Bioelectron.* **2019**, *148*, 111815. [CrossRef]
41. Huang, H.; Bai, J.; Li, J.; Lei, L.L.; Zhang, W.J.; Yan, S.J.; Li, Y.X. Fluorometric and colorimetric analysis of alkaline phosphatase activity based on a nucleotide coordinated copper ion mimicking polyphenol oxidase. *J. Mater. Chem. B* **2019**, *7*, 6508–6514. [CrossRef] [PubMed]
42. Mei, Y.Q.; Hu, Q.; Zhou, B.J.; Zhang, Y.H.; He, M.H.; Xu, T.; Li, F.; Kong, J.M. Fluorescence Quenching Based Alkaline Phosphatase Activity Detection. *Talanta* **2017**, *176*, 52–58. [CrossRef] [PubMed]

Review

Progress and Prospects of Electrochemiluminescence Biosensors Based on Porous Nanomaterials

Chenchen Li [1,2], Jinghui Yang [2], Rui Xu [2], Huan Wang [1,*], Yong Zhang [2,*] and Qin Wei [1]

[1] Collaborative Innovation Center for Green Chemical Manufacturing and Accurate Detection, Key Laboratory of Interfacial Reaction & Sensing Analysis in Universities of Shandong, School of Chemistry and Chemical Engineering, University of Jinan, Jinan 250022, China; lichenchen1107@126.com (C.L.); sdjndxwq@163.com (Q.W.)

[2] Provincial Key Laboratory of Rural Energy Engineering in Yunnan, Yunnan Normal University, Kunming 650500, China; ynnu_yangjh2002@126.com (J.Y.); ecowatch_xr@163.com (R.X.)

[*] Correspondence: wanghuan8711@163.com (H.W.); yongzhang7805@126.com (Y.Z.)

Citation: Li, C.; Yang, J.; Xu, R.; Wang, H.; Zhang, Y.; Wei, Q. Progress and Prospects of Electrochemiluminescence Biosensors Based on Porous Nanomaterials. *Biosensors* **2022**, *12*, 508. https://doi.org/10.3390/bios12070508

Received: 14 June 2022
Accepted: 7 July 2022
Published: 11 July 2022

Publisher's Note: MDPI stays neutral with regard to jurisdictional claims in published maps and institutional affiliations.

Copyright: © 2022 by the authors. Licensee MDPI, Basel, Switzerland. This article is an open access article distributed under the terms and conditions of the Creative Commons Attribution (CC BY) license (https://creativecommons.org/licenses/by/4.0/).

Abstract: Porous nanomaterials have attracted much attention in the field of electrochemiluminescence (ECL) analysis research because of their large specific surface area, high porosity, possession of multiple functional groups, and ease of modification. Porous nanomaterials can not only serve as good carriers for loading ECL luminophores to prepare nanomaterials with excellent luminescence properties, but they also have a good electrical conductivity to facilitate charge transfer and substance exchange between electrode surfaces and solutions. In particular, some porous nanomaterials with special functional groups or centered on metals even possess excellent catalytic properties that can enhance the ECL response of the system. ECL composites prepared based on porous nanomaterials have a wide range of applications in the field of ECL biosensors due to their extraordinary ECL response. In this paper, we reviewed recent research advances in various porous nanomaterials commonly used to fabricate ECL biosensors, such as ordered mesoporous silica (OMS), metal–organic frameworks (MOFs), covalent organic frameworks (COFs) and metal–polydopamine frameworks (MPFs). Their applications in the detection of heavy metal ions, small molecules, proteins and nucleic acids are also summarized. The challenges and prospects of constructing ECL biosensors based on porous nanomaterials are further discussed. We hope that this review will provide the reader with a comprehensive understanding of the development of porous nanomaterial-based ECL systems in analytical biosensors and materials science.

Keywords: porous nanomaterials; electrochemiluminescence; metal-organic frameworks; covalent organic frameworks; metal-polydopamine frameworks; biosensors

1. Introduction

Among various electroanalysis techniques, the electrochemiluminescence (ECL) method has attracted widespread attention due to its merits of high sensitivity and excellent analytical performance [1–3]. In the mid-1960s, Hercules, Santhanam, and Bard reported the first study on ECL [4,5]. Since then, the ECL technology has received wide attention from the scientific community. The ECL is a kind of new assay that combines two analytical methods: chemiluminescence (CL) and electrochemical techniques. Different from the traditional CL assay, ECL does not require an external excitation light source, so it has the advantages of a wider linear detection range, a better repeatability and anti-interference, a small background value, a good accuracy, and a high sensitivity [6,7]. The equipment and instruments for fabricating an ECL platform are usually small, and the preparation process is relatively simple and controllable [8,9]. Furthermore, ECL analysis can achieve continuous measurement, which is very popular and appropriate in the field of biochemical analysis, immunoassay, and pharmaceutical analysis. Attributing to these unique advantages, it has become one of the most highly interesting research areas for researchers in the field of analytical chemistry and been regarded as a very promising analytical assay [10–12].

With the inherent merits of a large specific surface area, a high porosity, an adjustable pore size and structure, and easy modification, porous nanomaterials have great potential in the fields of multiphase catalysis, gas adsorption and separation, drug transport, and biosensing [13–16]. In the field of ECL research, it is interesting that porous nanomaterials can not only be used as carriers for loaded luminophores, catalysts for accelerating the decomposition of catalytic co-reactants, and nanoreactors for accommodating ECL systems, but also accelerate the process of substance transport and charge transfer, all of which make ECL systems based on porous nanomaterials have a strong ECL response and a high sensitivity [17–21].

Sensitivity, stability, and reproducibility are important indicators for ECL biosensors. In order to improve the performance of ECL biosensors, it is particularly important to find and develop luminophores with strong ECL signals and a high stability. Conventional luminophores, such as g-C_3N_4, have an excellent ECL response, but their sheet-like structure makes their specific surface area relatively small, which reduces the probability of contact between the luminophore and the co-reactant and seriously affects their ECL luminescence efficiency. Luminol, tris-2-2′-bipyridyl ruthenium (i.e., Ru(bpy)$_3^{2+}$), and their derivatives possess good water solubility, making it difficult to apply them alone as luminophores in aqueous solutions. Therefore, finding nanomaterials with a large specific surface area and a high porosity as carriers of ECL luminophores or preparing ECL materials with a large specific surface area and a high porosity is of extraordinary significance for the preparation of high-performance ECL biosensors. In recent years, porous nanomaterials, such as ordered mesoporous silica (OMS) [22], metal–organic frameworks (MOFs) [23], covalent organic frameworks (COFs) [24], and metal–polydopamine frameworks (MPFs) [25] have received extensive attention in the study of ECL biosensors because of their large specific surface area, high porosity, tunable pore size and structure, and easy modification. It has also demonstrated excellent performance in the detection of heavy metal ions, small molecules, proteins, and nucleic acids.

In this review, we present a detailed description of porous nanomaterial-based ECL biosensors, combining the basic construction process and the applied reaction mechanism to show the innovative applications of porous nanomaterials in ECL biosensors, as shown in Figure 1. We further discuss the challenges and the prospects of ECL systems based on porous nanomaterials. This review will enable readers to understand the relevant contents comprehensively and find more innovative applications.

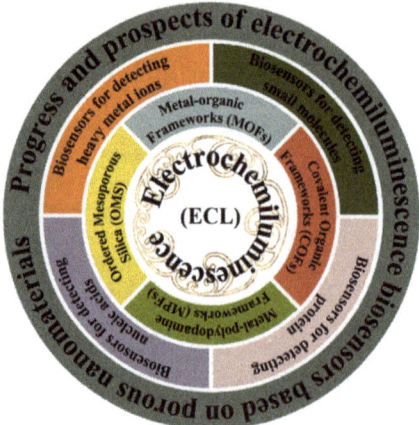

Figure 1. Schematics illustrating the application of ECL biosensors based on porous nanomaterials, as reviewed in this paper.

2. Synthesis of Porous Nanomaterials with ECL Properties

Numerous experimental results have shown that ECL biosensors based on porous nanomaterials can effectively enhance the ECL performance and enhance the accuracy of analysis [26,27]. Therefore, we firstly summarize and discuss the preparation of those porous nanomaterials with ECL properties, such as OMS, MOFs, COFs, and MPFs.

2.1. Ordered Mesoporous Silica (OMS) with ECL Properties

OMS has a wide range of applications in ECL biosensors, because of its excellent morphological characteristics, excellent stability, and simple preparation method. It is usually used as a carrier to load ECL substances by means of doping or coating techniques [28,29]. To make the discussion clearly, Table 1 summarizes several kinds of OMS-based nanocomposites being applied in the field of ECL biosensors and the corresponding synthesis strategies in recent years.

Table 1. Different synthetic strategies for OMS-based nanocomposites with ECL properties.

Nanocomposites	Methods	Luminous Body	Duration	Ref.
mSiO$_2$@CdTe@SiO$_2$ NSs	In situ synthesis	CdTe QDs	Microemulsion method	[30]
g-C$_3$N$_4$@ms-SiO$_2$	Post-synthesis modification	g-C$_3$N$_4$	Agitating	[31]
Ru-QDs@SiO$_2$	In situ synthesis	CN QDs, Ru(bpy)$_3^{2+}$	Microemulsion method	[32]
Ru@SiO$_2$	In situ synthesis	Ru(bpy)$_3^{2+}$	Self-assembly	[33]
CdTe@SiO$_2$	In situ synthesis	CdTe QDs	Microemulsion method	[34]
NH$_2$–Ru@SiO$_2$-NGQDs	Post-synthesis modification	CNQDs, Ru(bpy)$_3^{2+}$	Agitating	[35]
Ru@SiO$_2$ NPs	In situ synthesis	Ru(bpy)$_3^{2+}$	Microemulsion method	[36]
SiO$_2$@Ir	In situ synthesis	Ir(ppy)$_3^{2+}$	Microemulsion method	[37]
SiO$_2$@CQDs/AuNPs/MPBA	Post-synthesis modification	C QDs	Agitating	[38]
Ru@SiO$_2$	Post-synthesis modification	Ru(bpy)$_3^{2+}$	Agitating	[39]
SiO$_2$@Ru-NGQDs	In situ synthesis	Ru(bpy)$_3^{2+}$	Microemulsion method	[40]

As shown in Table 1, the methods for the synthesis of OMS with ECL properties are broadly divided into two categories. One is to synthesize silica nanoparticles (SiO$_2$ NPs) and obtain OMS-based nanocomposites through sodium hydroxide (NaOH) etching on this basis. The other is to encapsulate small organic particles in SiO$_2$ NPs by the microemulsion method and then obtain OMS-based nanocomposites by the high temperature calcination method [41]. The difficulty in controlling the process of etching SiO$_2$ NPs by NaOH makes it hard to get OMS with uniform pore channels by this etching method. Recently, the OMS prepared through the microemulsion method can effectively solve such problems [30,31]. For example, Lin et al. prepared OMS using this method, in which the luminescent g-C$_3$N$_4$ was combined with the previously prepared OMS by post-modification method as an efficient ECL probe. Based on this, the prepared sensor showed excellent correlation in the range of 0.1 nm–10 µm with an extremely low limit of detection (LOD) of 33 pM.

In another work, You et al. directly employed a microemulsion method to encapsulate Ru(bpy)$_3^{2+}$ and CN QDs in SiO$_2$ NPs. The electrons were transferred from CN QDs to Ru(bpy)$_3^{2+}$ through an intramolecular pathway, which shortened the distance between the electron transfer and thus improved the luminescence efficiency, yielding a self-enhanced ECL signal probe [32]. In a creative study, Jin et al. prepared a homogeneous Ru@SiO$_2$ NP colloidal solution and then applied it to develop Ru@SiO$_2$ NP nanomembranes on the surface of indium tin oxide glass (ITO) by a liquid–liquid interface self-assembly method. The obtained Ru@SiO$_2$ NP nanomembrane can be used as both an enhanced substrate and a luminol enricher. A self-enhanced ECL biosensor was constructed based on the intense luminescence of the Ru@SiO$_2$ NP nanomembrane and the enrichment of Ru(bpy)$_3^{2+}$

molecules on the surface of the Ru@SiO$_2$ NP nanomembrane [33]. In order to study the effect of the preparation process of nanocomposites on the properties of ECL, Shen et al. prepared CdTe@SiO$_2$ and SiO$_2$@CdTe NPs via microemulsion and post-modification methods, respectively. Interestingly, CdTe@SiO$_2$ with ordered mesopores is more efficient and less bio toxic for the preparation of ECL biosensors. The ECL immunosensor for the detection of methemoglobin was prepared using CdTe@SiO$_2$ as the signal probe with a good linearity in the range of 1.0 pg/mL to 100 ng/mL and the LOD was 0.22 pg/mL [34].

2.2. Metal–Organic Frameworks (MOFs) with ECL Properties

MOFs are a new type of porous material formed by organic ligands and metal ions or metal clusters linked by coordination bonds. Due to it inherit merits, such as a large surface area, a high porosity, abundant active sites, and a strong mass transfer capability, it has excellent performance in the field of novel materials and has a wider application in the fields of multiphase catalysis, gas adsorption and separation, drug transport, and biosensing [42–45]. Table 2 summarizes recent reported MOFs used in the field of ECL biosensors and the synthesis strategies of their composite, especially illustrating post-synthetic modifications, in situ synthesis and self-luminescent MOFs.

Table 2. Summary of different synthetic strategies for MOF composites with ECL properties.

MOF Composites	Ligands	Metal Source	Ref.
In situ synthesis			
MIL-101(Al)–NH$_2$	NH$_2$-BDC	AlCl$_3$	[46]
IRMOF-3	NH$_2$-BDC	Zn(NO$_3$)$_2$	[47]
Ru(bpy)$_3^{2+}$/NH$_2$-UiO-66	NH$_2$-BDC	ZrCl$_4$	[48]
Fe(III)-MIL-88B-NH$_2$	NH$_2$-BDC	FeCl$_3$	[49]
UiO-67	BPDC	ZrCl$_4$	[50]
GSH-Au NCS@ZIF-8	2-MI	Zn(NO$_3$)$_2$	[51]
Zinc Oxalate MOFs	Oxalic acid	Zn(NO$_3$)$_2$	[52]
Post-synthesis modifications			
Ru-MOF-5 NFs	PTA	Zn(NO$_3$)$_2$	[53]
Cu/Co-MOF	2-MI	Co(NO$_3$)$_2$, Cu(NO$_3$)$_2$	[54]
HH-Ru-UiO66-NH$_2$	NH$_2$-BDC	ZrCl$_4$	[55]
Co-Ni/MOF	2-MI	Co(NO$_3$)$_2$, Ni(NO$_3$)$_2$	[56]
AgNPs@Ru-MOF	NH$_2$-BDC	ZrCl$_4$	[57]
g-C$_3$N$_4$@NH$_2$-MIL-101	NH$_2$-BDC	FeCl$_3$·6H$_2$O	[58]
Zn-Bp-MOFs	H3BTC,4,4-dipyridyl	Zn(NO$_3$)$_2$	[59]
Ru-PCN-777	H$_3$TATB	ZrOCl$_2$	[60]
Self-luminous MOFs			
Eu-MOFs	5-bop	EuCl$_3$	[61]
RuMOF NS	[Ru(H$_2$dcbpy)$_3$]Cl$_2$	Zn(NO$_3$)$_2$	[62]
Eu-MOF	[Ru(H$_2$dcbpy)$_3$]Cl$_2$	Eu(NO$_3$)$_3$	[63]
Zr-TCBPE-MOF	H$_4$TCBPE	ZrCl$_4$	[64]
Hf-TCBPE	H$_4$TCBPE	HfCl$_4$	[65]
Zr$_{12}$-adb	H$_2$adb	ZrCl$_4$	[66]
Tb-Cu-PA MOF	IPA	TbCl$_3$, Cu(NO$_3$)$_2$	[67]
Zn-PTC	PTC	Zn(CH$_3$COO)$_2$	[68]
Ru@Zr$_{12}$-BPDC	BPDC, H$_2$dcbpy	ZrCl$_4$	[69]
Cu:Tb-MOF	IPA	TbCl$_3$, Cu(NO$_3$)$_2$	[70]
UMV-Ce-MOF	H$_3$BTC	Ce(NO$_3$)$_3$	[71]
PTP/Eu MOF	H$_3$BTC	Eu(NO$_3$)$_3$	[72]
Ce-TCPP-LMOF	TCPP	Ce(NO$_3$)$_3$	[73]
Zn-MOF	Hcptpy	ZnSO$_4$	[74]

2.2.1. In Situ Synthesis

Although MOFs have a high porosity and a tunable pore size, the pore size could be fixed along with the successful preparation of MOFs. Most classical MOFs contain only micropores, which leads to the inability of guest luminophores to easily enter the MOF interior through the pores, making the luminophore loading capacity of MOF materials greatly be reduced [75]. To solve this problem, Yang et al. wrapped the luminescence inside the MOF material by an in situ synthesis method during the MOF growth process, which resulted in a greatly enhanced loading capacity of the guest luminescence, and thus prepared the nanocomposite with a high-intensity ECL response [47].

As we all know, as an excellent luminescent substance, $Ru(bpy)_3^{2+}$ is often used in the process of constructing various ECL biosensors. For instance, Cao et al. encapsulated $Ru(Bpy)_3^{2+}$ molecules into NH_2-UiO-66 through the ligand effect during the growth of NH_2-UiO-66. The open channels or active cavities of MOFs could not only maintain the excellent ECL response of $Ru(bpy)_3^{2+}$, but also enrich the co-reactants, enabling the ECL biosensor to exhibit a highly selective and efficient ECL response, thus facilitating the ECL biosensor for ultra-sensitive and accurate analyses of the β-amyloid [48]. As shown in Figure 2D, Wang et al. applied mesoporous, hollow MIL-101(Al)-NH_2 in an ECL system by the in situ growth process and achieved a large and stable loading of $Ru(bpy)_3^{2+}$. Additionally, the authors also made poly(ethylenediamine) as a co-reactant and combined it with MIL-101(Al)-NH_2 through covalent bonding, which not only prevented the leakage of $Ru(bpy)_3^{2+}$, but also made the Ru complex produce strong and stable ECL signals through self-enhancement effect [46].

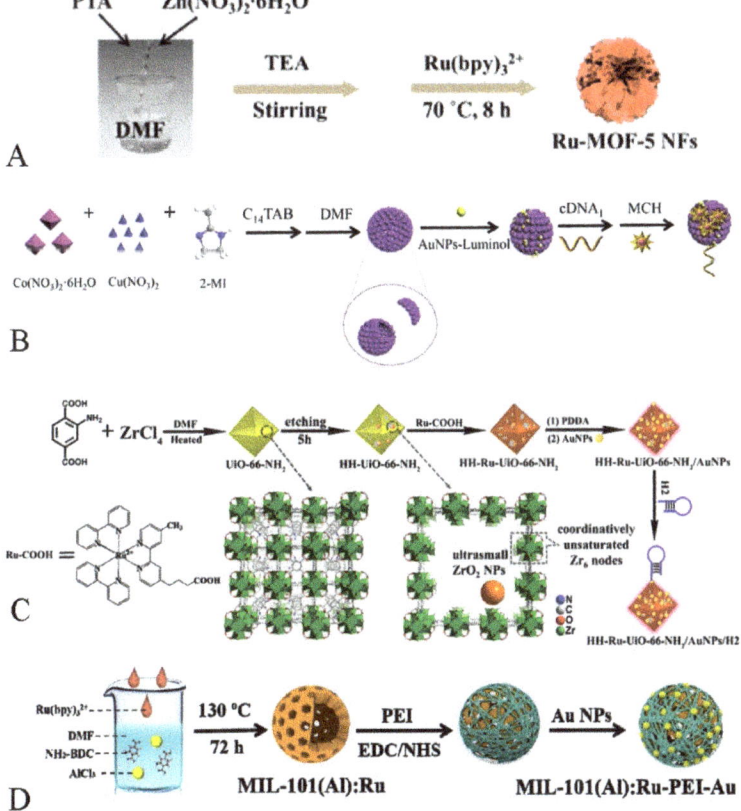

Figure 2. (**A**) The preparation of Ru-MOF-5 NFs [53]. Copyright © 2021, Elsevier. (**B**) The synthesis process of Cu/Co-MOF-luminol-AuNPs [54]. Copyright © 2021, Elsevier. (**C**) Preparation of HH-Ru-UiO-66-NH_2/Au NPs/H_2 [55]. Copyright © 2021, American Chemical Society. (**D**) The synthesis steps of MIL-101(Al): RuPEI-Au [46]. Copyright © 2019, Elsevier.

In addition to Ru(bpy)$_3^{2+}$, quantum dots (QDs) are often used in the field of ECL biosensors as an efficient and stable luminescent [76,77]. In the recent study, Tan et al. attached a large number of CdS QDs to chain-like polyethylenimine (PEI) via amide bonding, wrapped the modified PEI on the surface of MIL-53(Al), achieved massive loading of CdS QDs, and finally prepared MOF-based ECL signaling probes [78]. In another study, Deng et al. successfully encapsulated ZnSe QDs in Fe (III)-MIL-88B-NH$_2$ through the in situ growth process. Fe(III)-MIL-88B-NH$_2$ can not only achieve massive loading of ZnSe QDs, but it also contains amino groups for catalyzing the conversion of the co-reactant $S_2O_8^{2-}$ into sulfate anions ($SO_4^{\bullet -}$), which shortens the electron transfer distance and reduces the energy loss, enabling the ECL property of ZnSe QDs [49]. Although all of the above works demonstrate that the preparation of nanomaterials with ECL properties by the in situ growth method is an excellent strategy, the leakage of luminescent material is still an inevitable issue in the practical operations.

2.2.2. Post-Synthesis Modification

Post-synthesis modification is an effective means to prepare MOFs with ECL properties. The mechanism is mainly to attach luminophores to the surface of MOFs by ligand reaction, electrostatic force adsorption, or amide bonding, so that the MOFs materials, originally without ECL properties, become nanocomposites with ECL properties [79,80].

For example, As shown in Figure 2A, Wei et al. indicated that a flower-like nanomaterial with ECL properties was successfully prepared by loading Ru(bpy)$_3^{2+}$ onto the surface of MOF-5 NFs via electrostatic force adsorption, exhibiting an excellent ECL response [53].

It is worth noting that MOFs with catalytic properties act as carriers, not only to enrich the co-reactants but also to catalyze the co-reactants so that the ECL response is enhanced. As shown in Figure 2B, Zhou et al. successfully prepared nanocomposites with ECL properties by ligand bonding luminol to the MOF. The obtained hollow Cu/Co-MOF not only acted as a carrier but also could catalyze the generation of more O_2^- from H_2O_2, which greatly enhanced the ECL response by means of this material [54].

As shown in Figure 2C, Yuan et al. connected the luminescent material (Ru(Bpy)$_2$(Mcpbpy)$^{2+}$) with the carrier MOF (HH-UiO-66-NH$_2$) through amide bonding. On the one hand, the hierarchical pore shell and hollow cavity of HH-UiO-66-NH$_2$ exposed more amino groups, which made the loading of the luminescent greatly increased. On the other hand, the HH-UiO-66-NH$_2$ surface amino group could catalyze the generation of $SO_4^{\bullet -}$ from $S_2O_8^{2-}$, which greatly shortened the distance between the co-reactant and the luminophore, making the charge transfer more efficient and thus enhancing the ECL signals [55].

2.2.3. Self-Luminous MOFs

Luminous metal-organic frameworks (LMOFs) that consist of organic bridging ligands and metal-linked nodes are novel porous nanomaterials commonly used in ECL biosensors in recent years [81]. Since LMOFs have multiple structural units, the ECL response may come from metal centers and ligands within the MOF, and the optical properties can be modulated by the interactions between the building components. Based on this, the problems of low loading of modified luminescent substances and leakage from the in situ-grown luminescent substances after MOF synthesis might be effectively solved [62,82].

Lanthanide rare-earth metal ions are an important emerging LMOFs precursor because of their unique [Xe]4f^n (n = 0–14) ground-state electronic grouping pattern, which is prone to 4f–4f transitions and possesses abundant ladder electron energy levels and sharp emission bands. As shown in Figure 3, Wei et al. prepared self-luminous (Ln) metal-organic frameworks (Ln-MOFs) by hydrothermal treatment using Eu (III) ions and 5-boryl-isophthalic acid (5-bop) as precursors. The 5-bop in the triplet excited state can transfer its own energy to the Eu(III) ion by emitting ultraviolet light. When the Eu(III) ion gains energy from the ligand, it can jump to a higher energy level and release more light energy when it falls back to the ground state. The more intense ECL signal is obtained through the antenna effect of Eu(III) ions. The prepared sensor showed a good linearity in the range of 0.005 to 100 ng/mL, and the obtained LOD was only 0.126 pg/mL [61].

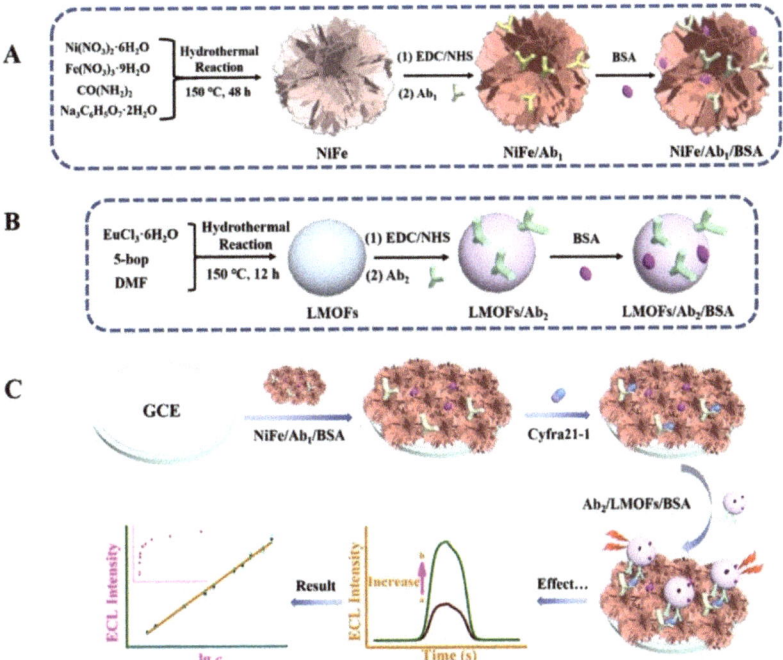

Figure 3. Preparation of NiFe Complex/Ab$_1$/BSA Bioconjugate (A), Ab$_2$/LMOFs/BSA Bioconjugate (B), and Formation Route of the Suggested Signal-Enhanced ECL Model (C) [61]. Copyright © 2021, American Chemical Society.

Ru(bpy)$_3^{2+}$ and its derivatives can not only prepare porous nanomaterials with ECL properties by post-synthesis modification or in situ growth, but also act as ligands for the direct synthesis of self-luminous MOFs involved in the construction of ECL biosensors [73,83].

In a pioneering work conducted by Yan et al., self-luminous Ru-MOF has been synthesized by using the autoloading Ru(dcbpy)$_3^{2+}$ and zinc ions as precursors. The obtained Ru-MOF nanosheets expose more active centers, promote closer contact with the target molecule, and have shorter diffusion distances for ions, electrons, and co-reactants. The excellent property makes the self-luminous Ru-MOF show great potential as a new Faraday cage for developing a biosensing platform [62].

As shown in Figure 4, in another interesting work, Zhao et al. synthesized a new type of Eu-MOF by a hydrothermal method using Eu (III) ions and Ru(dcbpy)$_3^{2+}$ as precursors. The Eu-MOF can undergo redox reactions and energy transfer between its ligand molecules and achieve annihilation luminescence without any additional co-reactants. At the same time, the antenna effect of Eu (III) ions in Eu-MOF is generated. In other words, when Eu (III) ions absorb the energy from the ligand, the luminescence efficiency is greatly increased, and the secondary near-infrared (NIR-II) luminescence is obtained. The prepared ECL sensor using Eu$_2$[Ru(Dcbpy)$_3$]$_3$ as the ECL signal probe was extremely resistant to interference and achieved the rapid and sensitive detection of trenbolone in the range of 5 fg/Ml–100 ng/Ml with a lower LOD of 4.83 fg/Ml [63].

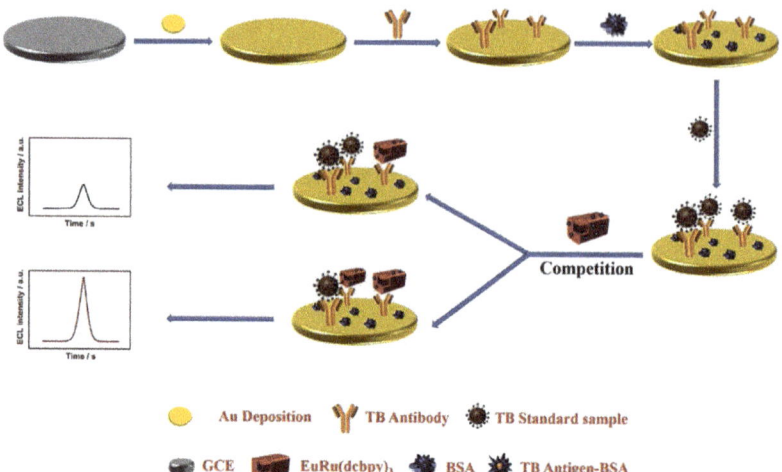

Figure 4. Annihilation luminescent Eu-MOF as a near-infrared electrochemiluminescence probe for trace detection of trenbolone [63]. Copyright © 2022, Elsevier.

Although Ru(bpy)$_3^{2+}$ and its derivatives have very common applications in the field of ECL, their high price and the biotoxicity carried by the co-reactants limit the current application of Ru-containing MOFs.

In particular, aggregation-induced ECL (AI-ECL) was firstly discovered by Luisa De Cola's team in 2017 [84]. With the continuous development in recent years, some excellent AI-ECL materials have emerged. Recently, researchers found that tetraphenylethylene (TPE) and its derivatives have the characteristics of a high ECL efficiency via easy modification. Compared with its aggregates and monomers, MOFs prepared based on TPE showed a stronger ECL response. For instance, Yuan et al. successfully prepared a novel 2D ultrathin MOF material based on the aggregation-induced emission (AIE) ligand H4ETTC and used it to construct a novel ECL biosensor for the ultrasensitive detection of CEA. The newly synthesized AIE luminogen (AIEgen)-based MOF (Hf-ETTC-MOF) yielded a higher ECL intensity and efficiency than H$_4$ETTC monomers, H$_4$ETTC aggregates and 3D bulk Hf-ETTC-MOF did [85]. As shown in Figure 5, Wei et al. synthesized a dumbbell-shaped metal–organic backbone with high luminescence efficiency by combining the aggregation-induced luminescent material H$_4$TCBPE with Zr(IV) ions. The obtained Zr-TCBPE-MOF possesses a more excellent ECL performance compared to the monomer and aggregates of H$_4$TCBPE. In addition, the authors combined Zr-TCBPE-MOF with polyethyleneimine (PEI) to prepare a unique self-reinforced Zr-TCBPE-PEI electroluminescent complex, which could effectively avoid the bio-toxicity of the co-reactant and exhibit a more dramatic ECL response. The prepared ECL sensor showed good correlation in the range of 0.0001–10 ng/mL and the lower LOD was 52 fg/mL, providing an effective way for the early and sensitive detection of small cell lung cancer [64].

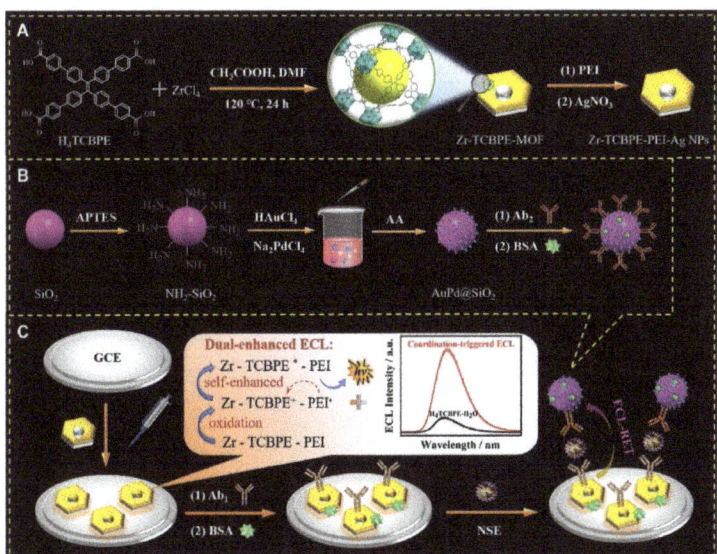

Figure 5. Schematic illustration for the synthesis of (**A**) Zr-TCBPE-PEI-Ag NPs substrate and (**B**) Ab$_2$-AuPd@SiO$_2$ bioconjugate, and (**C**) the fabrication process of the proposed ECL immunosensor with possible ECL-enhancing effects and luminescence mechanism [64]. Copyright © 2022, John Wiley and Sons.

2.3. Covalent Organic Frameworks (COFs) with ECL Properties

COFs is a class of organic porous crystalline materials composed of light elements (C, O, N, B, etc.), which are connected to each other by covalent bonds. Due to their structural designability, low density, high specific surface area, easy modification, and functionalization, COFs have been widely investigated and shown excellent prospects for applications in the fields of gas storage and separation, non-homogeneous catalysis, energy storage materials, optoelectronics, sensing, and drug delivery [86].

Although COFs materials currently synthesized by solvothermal synthesis and Knoevenagel polycondensation reaction are often used for ECL biosensors, there are not so many articles reported about the application of COFs in the development of ECL biosensors. Recently, Zhuo et al. prepared a nanocomposite with ECL properties by attaching Ru(bpy)$_3^{2+}$ to the surface of COF-LZU1. Since COF-LZU1 has a hydrophobic porous structure and TPrA is lipophilic, a large amount of TPrA in aqueous solution can be enriched into the hydrophobic inner cavity of COF-LZU1, which will shorten the distance between the luminescent material and the co-reactant and increase the concentration of co-reactant around the luminescent, resulting in a greatly enhanced ECL response [87]. However, the conductivity of COFs materials limits its application in the ECL field, which may be one of the reasons for limiting applications of COFs in the ECL biosensing study. To solve this problem, Yuan et al. provided a novel strategy. They prepared a conductive COF (HHTP-HATPCOF), as shown in Figure 6. In their work, since HHTP-HATP-COF has a large amount of ECL-emitting material and its conductive porous backbone accelerates the charge transfer in the whole backbone, the ECL response of this composite is greatly enhanced [88].

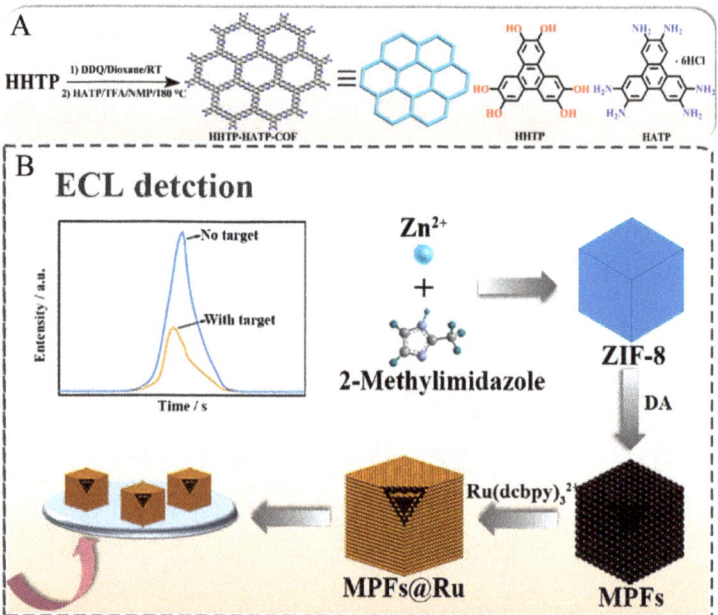

Figure 6. (**A**) Synthesis of HHTP-HATP-COF [88]. Copyright © 2022 American Chemical Society. (**B**) Preparation of MPFs@Ru [25]. Copyright © 2021 Elsevier.

2.4. Metal–Polydopamine Frameworks (MPFs) with ECL Properties

MPFs are a new hybrid material that perfectly combine the advantages of both MOFs and polydopamine (PDA) [89]. The large specific surface area and high porosity can be used to obtain a strong ECL signal by increasing the loading of luminophores. The PDA structure contains active double bonds that can chemically react with multiple groups to connect luminophores. The PDA structural fragment is a conjugated system that can generate π–π stacking with luminophores containing π bonds and thus adsorb luminophores [90,91].

As shown in Figure 6, Ma et al. prepared the MOF (ZIF-8) as the basic framework by the self-assembly method at first. After that, the hollow and porous metal–polydopamine frameworks (MPFs) were gradually formed by reacting with dopamine in a mixture of ethanol and Tris-HCl buffer, in which the polydopamine continuously replaced the original ligands through coordination reactions. Finally, the nanocomposites with ECL properties were formed by adsorption of Ru(bpy)$_3^{2+}$ through π–π stacking [25]. Although MPFs have many advantages, there are still less related studies on applying MPFs in preparing ECL biosensors. So, it is hoped that MPFs can achieve greater breakthroughs in the field of ECL biosensors through the continuous efforts of researchers.

3. Application of ECL Biosensors Based on Porous Nanomaterials

Due to their excellent stability and selectivity, porous nanomaterials have a wide range of applications in the field of ECL biosensors [92,93]. Here, we focus on introducing the applications of ECL biosensors based on porous nanomaterials in the detection of heavy metal ions, small molecules, proteins, and nucleic acids in recent years.

3.1. Biosensors for Detecting Heavy Metal Ions

It is well known that heavy metal ions have a great impact on human health and the natural environment, especially the intake of large amounts of heavy metal ions can cause irreversible damage to the human body, so the reliable and accurate detection of heavy metal ions is of great importance [94]. For example, You et al. prepared a label-free ECL biosensor for the detection of Hg^{2+} based on the different affinity of Ru-QDs@SiO$_2$ nanocomposites for loading single-stranded DNA (SsDNA) and Hg^{2+}-initiated double-stranded DNA (DsDNA). When no ions of Hg^{2+} are present, single-stranded DNA is attached to the Ru-QDs@SiO$_2$ surface by hydrogen bonding, i.e., electrostatic force adsorption, which leads to the quenching of the ECL signal. When Hg^{2+} is present, part of the single-stranded DNA on the surface of Ru-QDs@SiO$_2$ is guided to form a stable dsDNA, allowing part of Ru-QDs@SiO$_2$ to exist in a free state, reducing the quenching of the ECL signal to single-stranded DNA [32]. Additionally, You et al. attached one end of the Hg^{2+} aptamer to NH$_2$-Ru@SiO$_2$-NGQds through an amide bond and the other end to AuNPs on the surface of the glassy carbon electrode (GCE) through an Au–S bond. When Hg^{2+} is absent, the aptamer is a long chain, and there is a large spatial site resistance between the luminescent material and the electrode surface, and, thus, the ECL response is not strong. When Hg^{2+} is present, the aptamer bends due to the formation of a thymine-Hg^{2+}-thymine (T-Hg^{2+}-T) specific structure, which draws the distance between the luminescent material and the electrode surface and reduces the spatial potential resistance, making the ECL response greatly enhanced [35].

3.2. Biosensors for Detecting Small Molecules

Table 3 summarizes ECL biosensors for detecting the small molecule based on porous nanocomposites in recent years.

Table 3. Summary of the construction of ECL biosensors based on porous nanocomposites for small molecule detection.

Analytes	Nanocomposites	Linear Range	LOD	Ref.
DES	Ru@SiO2	4.8×10^{-4}~36.0 nM	0.025 pM	[39]
DES	UiO-67	0.01 pg/mL~50 ng/mL	3.27 fg/mL	[50]
Rutin	GSH-Au NCS@ZIF-8	0.05~100 μM	10 nM	[51]
Acetamiprid	Cu/Co-MOF	0.1 μM~0.1 pM	0.018 pM	[54]
CAP	Co-Ni/MOF	1.0×10^{-13}~1.0×10^{-6} M	2.9×10^{-14} M	[56]
ATX-a	AgNPs@Ru-MOF	0.001~1 mg/mL	0.00034 mg/mL	[57]
Trenbolone	Eu-MOF	5 fg/mL~100 ng/mL	4.83 fg/mL	[63]
IMI	UMV-Ce-MOF	2–120 nM	0.34 nM	[71]
Lincomycin	PTP/Eu MOF	0.1 mg/mL~0.1 ng/mL	0.026 ng/mL	[72]

Chloramphenicol (CAP) is a broad-spectrum antibiotic that can effectively treat a variety of microbial infections such as typhoid fever, meningitis, and salmonellosis. Since the middle of last century, it has become a widely used antibiotic because of its low production cost and good drug stability. However, many studies in recent years have shown that excessive intake of CAP can inhibit bone marrow hematopoiesis, which in turn can severely damage the human hematopoietic system. Therefore, it is necessary to establish a rapid and sensitive detection method to accurately monitor CAP residues in the aqueous environment. Chen et al. synthesized black phosphorus QDs (BPQDs) into PTC-NH$_2$ solution to synthesize nanocomposites with ECL properties (BP/PTC-NH$_2$). An efficient and sensitive ECL sensor was prepared to detect CAP by combining BP/PTC-NH$_2$ with Co-Ni/MOF as an ECL emitter via electrostatic adsorption. The composite material of co-Ni/MOF has a good catalytic effect and can catalyze the co-reactant $K_2S_2O_8$ to generate more $SO_4^{\bullet-}$, which enhances the ECL response of the system. When CAP is present, the specific recognition of the aptamer makes the aptamer detach from the surface of the luminescent material, which reduces the burst of the aptamer for the ECL response and thus enhances the ECL signal [56].

Anatoxin-a (ATX-a) is a highly toxic alkaloid neurotoxin isolated from Anabaena flos-aquae (e.g., a semi-lethal dose of 200 μg/mL for rats) with a strong nicotine-like neuromuscular depolarization blocking effect, and animals poisoned will experience myofascicular twitching, corneal inversions, respiratory muscle spasms, and the animals will show symptoms such as muscle bundle convulsion, corkscrew, respiratory muscle spasm, and salivation after poisoning. Therefore, it is of great interest to find a rapid and sensitive test for the detection of ATX-a in water. As shown in Figure 7, Wang et al. proposed an ECL biosensor based on the ECL-RET strategy with a low background signal by means of double burst and dual stimulus response. The prepared ECL biosensor provides an accurate signal output for the ultrasensitive detection of ATX-a. Specifically, the authors wrapped Ru(bpy)$_3^{2+}$ in UiO-66-NH$_2$ by the in situ growth method to act as an ECL signal probe, wrapped it with silver nanoparticle (AgNPs) shells as the main bursting agent, and tightly bound it to DNA-ferrocene (Fc). The AgNPs play an important role in the whole system, not only to close the permanent pore of UiO-66-NH$_2$ and prevent the leakage of Ru(bpy)$_3^{2+}$, but also to act as a quencher to quench the ECL signal of Ru(bpy)$_3^{2+}$. Not only that, the AgNPs generated in situ can specifically recognize and break the substrate chain, generating an "On" signal, which helps to avoid false positive results. Thus, an ultra-sensitive detection of ATX-a was achieved in the range of 0.001 to 1 mg/mL, and the LOD was estimated to be 0.00034 mg/mL [57].

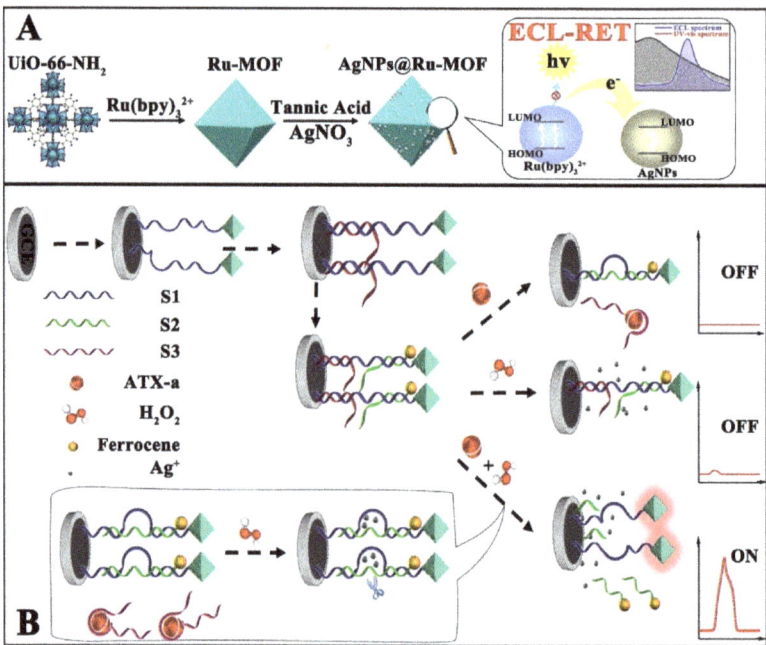

Figure 7. Schematic illustration of in situ generation of AgNPs@Ru-MOF and progress of ECL-RET (**A**), and the response process of ECL-RET aptasensor for ATX-a (**B**) [57]. Copyright © 2021, American Chemical Society.

3.3. Biosensors for Detecting Protein

Cancer has now become a global difficult-to-treat disease, and it is well known that the treatment of early-stage cancer patients saves much more labor, money, and time than that of late-stage cancer patients, so early diagnosis of cancer is of great significance. The ECL analysis method provide a potential assay for the sensitive and selective detection of certain cancer disease-related biomarkers by coupling antigen–antibody specific binding with ECL technology. As shown in Table 4, we summarize some of the porous nanomaterial-based ECL sensors used for protein detection in recent years.

Table 4. Summary of the construction of ECL biosensors based on porous nanocomposites for protein detection.

Analytes	Nanocomposites	Linear Range	LOD	Ref.
HE4	g-C_3N_4@ms-SiO_2	10^{-5} to 10 ng/mL	3.3×10^{-6} ng/mL	[31]
PSA	Ru@SiO_2	10^{-15} to 10^{-6} g/mL	0.169 fg/mL	[33]
AFP	CdTe@SiO_2	1.0 pg/mL to 100 ng/mL	0.22 pg/mL	[34]
HAase	Ru@SiO_2 NPs	2 to 60 U/mL	2 U/mL	[36]
BNPT	SiO_2@Ir	0.1 ng/mL to 200 ng/mL	0.03 ng/mL	[37]
AFP	SiO_2@CQDs/AuNPs/MPBA	0.001 to 1000 ng/m L	0.0004 ng/mL	[38]
PCT	MIL-101(Al)–NH_2	0.0005 ng/mL to 100 ng /mL	0.18 pg/mL	[46]
cTnI	IRMOF-3	1 fg/mL to 10 ng/mL	0.46 fg/mL	[47]
SCCA	Fe(III)-MIL-88B-NH_2	0.0001 to 100 ng/mL	31 fg/mL	[49]
Aβ	Zinc Oxalate MOFs	100 fg/mL to 50 ng/mL	13.8 fg/mL	[52]
NSE	Ru-MOF-5 NFs	0.0001 ng/mL to 200 ng/mL	0.041 pg/mL	[53]
Thrombin	HH-Ru-UiO66-NH_2	100 fM to 100 nM	31.6 fM	[55]
PCT	NH_2-MIL-101	0.014 pg/mL to 40 ng/mL	3.4 fg/mL	[58]
MUC1	Zn-Bp-MOFs	1 pg/mL to 10 ng/mL	0.23 pg/mL	[59]
MUC1	Ru-PCN-777	100 fg/mL to 100 ng/mL	33.3 fg/mL	[60]
CYFRA21-1	Eu-MOFs	0.005 to 100 ng/mL	0.126 pg/mL	[61]
cTnI	RuMOFNSs	1 fg/mL to 10 ng/mL	0.48 fg/mL	[62]
NSE	Zr-TCBPE-MOF	0.0001 to 10 ng/mL	52 fg/mL	[64]
MUC1	Hf-TCBPE	1 fg/mL to 1 ng/mL	0.49 fg/mL	[65]
MUC1	Zr_{12}-adb	1 fg/mL to 100 ng/mL	100 fg/mL	[66]
CYFRA21-1	Tb-Cu-PA MOF	0.01 to 100 ng/mL	2.6 pg/mL	[67]
MUC1	Ru@Zr_{12}-BPDC	1 fg/mL to 10 ng/mL	0.14 fg/mL	[69]
ProGRP	Cu:Tb-MOF	1.0 pg/mL to 50 ng/mL	0.68 pg/mL	[70]

Mucin-1 (MUC1) is an important transmembrane glycoprotein that is considered an important biomarker for colon, breast, ovarian, and lung cancers. Recently, Yuan et al. prepared a novel multivacancy nanocomposite (Hf-TCBPE) with ECL properties based on the principle of matrix coordination-induced ECL (MCI-ECL) enhancement. The MOF constructed by Hf ions and TCBPE ligands has its internal spatial structure fixed, which restricts the intramolecular free motion of TCBPE and suppresses the nonradiative relaxation, and the high porosity of Hf-TCBPE enables both internal and external excitation of TCBPE, which greatly enhances its ECL response. An ECL biosensor for the detection of mucin 1 (MUC 1) was constructed by combining HF-TCBPE with a phosphate-terminated ferrocene (FC)-labeled hairpin DNA aptamer (FC-HP3) as a signal probe (HF-TCBPE/FC-HP3) with the aid of an exonuclease III (Exo III) cyclic amplification strategy [65]. Another novel piece of research is that Xiao et al. discovered that the ECL of the material could be enhanced by restricting intramolecular motion, and based on this principle, a two-dimensional ultrathin MOF (Zr_{12}-ABD) with AI-ECL properties was prepared. In 2D MOF, the ligand 9,10-anthracene dibenzoate is immobilized, which restricts its intramolecular motion and suppresses the energy loss due to spin, allowing more energy to be released in the form of light energy and significantly enhancing the ECL response. Meanwhile, the ultrathin multivacancy structure of 2D MOF not only allows more co-reactants to enter the interior of MOF, but also reduces the migration distance between the electrons, ions, and co-reactants due to the smaller spatial site resistance, which reduces the energy loss and further enhances the ECL response of Zr_{12}-ABD. A biosensor for the sensitive detection of mucin 1 was prepared by combining Zr_{12}-ABD nanomaterials with a bipedal walking molecular machine. The ECL signal decreased with an increasing concentration of MUC1 in the range of 1 fg/mL to 100 ng/mL, showing a good linearity, and the LOD of the prepared ECL sensor was only 0.25 fg/mL [66].

CYFRA21-1 is considered to be a tumor marker mainly used for the detection of lung cancer and is especially valuable for the diagnosis of non-small cell lung cancer (NSCLC). In a study, Wei et al. creatively prepared a rare earth (Ln) metal–organic backbone (LMOF) with ECL properties. Ln-MOF was prepared from a precursor containing Eu(III) ions and

5-boronic acid isophthalic acid (5-bop). The ligand 5-bop produces a triplet state upon UV excitation, which triggers the red light emission of Eu(III) ions and enhances the ECL response. The electron-deficient boric acid reduces the energy transfer efficiency from the triplet state of 5-bop to the Eu(III) ion, resulting in both being efficiently excited under a single excitation. In addition, the synthesized flower-like Ni/Fe composites (Ni/Fe 1:1) have more active centers, higher stability, and good electrical conductivity by gradually adjusting the atomic ratio of Ni/Fe. An ECL immunosensor for the highly sensitive detection of CYFRA21-1 was prepared using Ln-MOF as the ECL emitter and flower-like Ni/Fe composite as the substrate, and the prepared Eu-LMOF showed good performance characteristics in the ECL immunoassay with the LOD of 0.126 pg/mL [61].

Ju et al. synthesized a MOF with ECL properties (Tb-Cu-PA MOF) using luminescent Tb^{3+} and catalytic Cu^{2+} ions as metal linkers and isophthalic acid (PA) as a bridging ligand. The doping of Cu^{2+} significantly reduced the size of the MOF and produced a strong and stable ECL signal. Therefore, the authors prepared a novel ECL immunosensor for the sensitive detection of CYFRA21-1 by using the synthesized Tb-Cu-PA MOF as an ECL emitter and Ni–Co layered double hydroxide (LDH) containing ZIF-67 nanocubes as a substrate. Compared with ZIF-67, ZIF-67@LDH has larger specific surface area and more active centers. After depositing palladium nanoparticles (Pt NPs) on ZIF-67@LDH nanocubes, it can improve the charge transport and electrocatalytic performance, catalyze $S_2O_8^{2-}$ to produce more $SO_4^{\bullet-}$, and obtain more intense ECL signals. The linear range of the successfully prepared ECL immunosensor was 0.01–100 ng/mL with a LOD of 2.6 pg/mL [67].

Except for the biomarkers, biological enzymes also play an irreplaceable role in human life activities. The abnormal activity of certain enzymes can lead to disorders in human functions. Therefore, to develop a rapid and highly sensitive measurement of some enzymes is important for clinical diagnosis and basic biochemical research. The sensitive detection of thrombin (TB), an important biomarker that plays an important role in hemostasis and thrombosis, has attracted great interest. In a study, Yuan et al. prepared a hollow graded MOF (HH-UiO-66-NH_2) with graded pore shells by a simple hydrothermal etching method and used it as a carrier to load $Ru(bpy)_2(Mcpbpy)^{2+}$, and successfully prepared a nanocomposite (HH-Ru-UiO66-NH_2) with an excellent ECL signal. The multilayer structure and cavity of HH-UIO-66-NH_2 allowed the macromolecule $Ru(bpy)_2(Mcpbpy)^{2+}$ to be immobilized not only on the surface of MOF but also on the interior of MOF, which led to a significant increase in the loading of MOF on the luminescent group. On the other hand, the multilayer structure of HH-UIO-66-NH_2 allowed the rapid diffusion of reactants, ions, and electrons, thus promoting the excitation of more luminophores. In addition, the etched HH-UIO-66-NH_2 exposes more amino groups, which can catalyze the co-reactant $K_2S_2O_8$ to generate $SO_4^{\bullet-}$ radicals, and thus greatly improve the ECL luminophore utilization rate. Ultimately, the authors used HH-RU-UIO-66-NH_2 as a high-performance ECL probe combined with a catalytic hairpin assembly (CHA) enzyme-free amplification technology to construct an ECL biosensor for the ultrasensitive detection of TB. The successfully prepared ECL immunosensor exhibited a good linearity in the range of 100 fM–100 nM, and the LOD of the prepared ECL sensor was only 31.6 fM [55].

Hyaluronidase (Haase) is another general term for enzymes that oligomerize hyaluronic acid (HA). It can decrease the activity of hyaluronic acid in the body, thereby increasing the ability of fluid permeation in tissues. In recent years, Haase has been considered as a potential tumor marker. In a recent work, Lin et al. designed a slow-release system based on a hydrogel constructed of HA with polyethyleneimine (PEI) and a large amount of $Ru(bpy)_3^{2+}$-doped silica nanoparticles (Ru@SiO_2NPs) stably dispersed in the hydrogel as an ECL signaling probe. When Haase is present, the hydrogel is decomposed by Haase, allowing Ru@SiO_2NPs to escape from the hydrogel into the supernatant, and the concentration of Haase can be quantified by the ECL signal generated from the supernatant. Compared with previous work, this biosensor does not require large amounts of HA to immobilize the signal probe or tedious centrifugation methods to reduce background interference, and thus has an excellent sensitivity and selectivity [36].

3.4. Biosensors for Detecting Nucleic Acids

MicroRNAs are small single-stranded non-coding RNAs of about 19–23 nucleotides. Although the first microRNAs were discovered as early as 1993, only in recent years has the diversity and breadth of this class of genes been revealed. It is hypothesized that vertebrate genomes have up to 1000 different miRNAs, regulating at least 30% of gene expression. Moreover, microRNAs are also considered to be reliable biomarkers for various cancers and genetic diseases. With the continuous development of biotechnology, scientists have combined DNA amplification strategies, such as target cyclic amplification (TRC), catalytic hairpin assembly (CHA), and strand displacement amplification (SDA). With the ECL technique to prepare a number of efficient and sensitive ECL biosensors for the detection of nucleic acids [95,96]. As shown in Table 5, an increasing number of ECL biosensors based on porous nanocomposites have been successfully developed and applied for the sensitive detection of nucleic acids in recent years.

Table 5. Summary of the construction of ECL biosensors based on porous nanocomposites for nucleic acids detection.

Analytes	Nanocomposites	Linear Range	LOD	Ref.
miRNA-182	mSiO$_2$@CdTe@SiO$_2$ NSs	0.1 to 100 pM	33 fM	[30]
microRNA-21	Zn-PTC	100 aM to 100 pM	29.5 aM	[68]
miRNA-133a	Zn-MOF	50 aM to 50 fM	35.8 aM	[74]
microRNA-21	Py-sp2c-COF	100 aM to 1 nM	46 aM	[97]
microRNA-21	Co-MOF-ABEI/Ti$_3$C$_2$T$_x$	0.00001 to 10 nM	3.7 fM	[98]
miRNA-155	RuMOFs	0.8 fM to 1.0 nM	0.3 fM	[99]

Recently, Yuan et al. found that when polycyclic aromatic hydrocarbons (PAHs) were used as ligands for the synthesis of MOFs, the aggregation-induced burst (ACQ) effect of PAHs could be effectively eliminated by coordination immobilization of the ligands as a way to improve the strength and efficiency of ECL. Based on this principle, a MOF (Zn-PTC) with AI-ECL properties was prepared. The molecular spacing of PTC was effectively increased by ligand immobilization in the MOF, thus eliminating the ACQ effect and resulting in a greatly enhanced ECL response of Zn-PTC. In addition, the PTCs in Zn-PTC stacked in an edge-to-edge manner to form J-type aggregates, which also promoted the enhanced ECL response. Based on the good ECL performance, an ECL biosensor for the sensitive detection of microRNA-21 was constructed by using Zn-PTC as an ECL signaling probe in combination with a dual amplification strategy of a nucleic acid exonuclease III-stimulated targeting cycle and DNAzyme-assisted cycling. The ECL sensor was able to detect microRNA-21 in the range of 100 aM to 100 pM with an efficient and sensitive LOD of 29.5 aM [68].

Additionally, Yuan's team prepared a Py-sp2 carbon-conjugated nanosheet (Py-sp2c-CON) with ECL properties based on the condensation reaction of tetrakis(4-formylphenyl)pyrene (TFPPy) and 2,2'-(1,4-phenylene)diacetonitrile. The porous ultra-thin structure of Py-sp2c-CON can effectively shorten the material transport distance and energy transfer process of electrons, ions, and co-reactants (S$_2$O$_8^{2-}$), which greatly enhances the ECL response of luminescent substances. Based on these advantages, an ECL biosensor for microRNA-21 detection was prepared using the Py-sp2c-CON/S$_2$O$_8^{2-}$/Bu$_4$NPF$_6$ system, which has a wide linear response (100 am~1 nM) and a lower LOD (46 am) [97].

In another study, a highly efficient and sensitive ECL biosensor for the detection of miRNA-21 was constructed using a Co-MOF-ABE/Ti$_3$C$_2$Tx composite as an ECL luminescent substance combined with a DSN-assisted target recovery signal amplification strategy. Co-MOF has a large specific surface area and thus can be loaded with abundant luminophores. Not only that, Co-MOF also exhibits excellent catalytic properties, and the ECL response of the composite is greatly improved by these factors. The successfully prepared ECL immunosensor achieved sensitive detection of miRNA-21 in the range of 0.00001 and 10 nM with a LOD of only 3.7 fM [98].

4. Conclusions and Outlooks

In one word, we give a comprehensive overview of the recent progress in the construction of ECL biosensors based on porous nanomaterials. Due to their large specific surface area, high porosity, large number of active sites, adjustable structure, easy modification, and good biocompatibility, porous nanomaterials, such as OMS, MOFs, COFs, and MPFs, have a large number of applications in the field of ECL biosensors. The preparation of different types of porous nanomaterials with ECL properties were summarized and their applications in the detection of heavy metal ions, small molecules, proteins, and nucleic acids. In conjunction with the representative articles, we summarize the advantages of porous nanomaterials in the field of ECL biosensors.

Firstly, porous nanomaterials with a large specific surface area and a variety of different functional groups can provide a good modification basis for loading ECL luminescent substances, which is beneficial for the construction of high-performance ECL biosensors.

Secondly, the high porosity and adjustable pore size can provide the basis for the preparation of composites by means of various synthesis methods, such as in situ growth and encapsulation, and it can additionally provide good channels for energy transfer and substance transport such as ions, electrons, and co-reactants.

Thirdly, the loading of co-reactants and luminescent substances in an all-in-one structure or the loading of two luminescent substances with different wavelengths in the same structure based on the RET effect can shorten the substance and charge transfer paths between the luminescent substances and basis, and thus can realize the self-enhancement of the ECL response, which reflects the superiority of the synergistic effect.

Although greater progress has been made in ECL biosensors based on porous nanomaterials, there are still some problems to be paid attention to that may limit the application of porous nanomaterials in this field. One of the main issues is that most porous nanomaterials with ECL properties in the field of macromolecular sensing still suffer from a high excitation voltage and a low luminescence efficiency because of their high impedance. Therefore, the development of luminescent substances with a high conductivity, a low voltage excitation, and a high ECL efficiency are also discussed in detail. Although porous nanomaterials have been widely used in the field of ECL biosensors, the limitations of synthesis methods and synthetic materials prevent the application of organic composite materials with a poor electrical conductivity (e.g., COFs materials) from being widely used. It is worth noting that the application of MPFs and AI-ECL materials are still in the initial stage, and so is the design of COFs materials with a high electrical conductivity. Based on this stage, the design of COFs with a high conductivity and the construction of ECL biosensors based on MPFs open new directions in the field of ECL.

1. AI-ECL materials or techniques are still a new research direction in the field of sensors. The types of ligands currently used are relatively single, and the AI-ECL mechanism of most materials is almost the same. Therefore, the search for new organic ligands, the study of a new AI-ECL material reaction mechanism, and the design of synthesizing innovative structures of AI-ECL materials will be hot directions.
2. Most of the current porous nanomaterials with functional groups have poor electrical conductivities, and the functional group types are relatively single. How to prepare porous nanomaterials with a high conductivity and multiple functional groups and how to combine them with luminescent substances with different functional groups to achieve the synergistic effect between each group are still needed to pay more attention to.
3. At present, most of the ECL biosensors based on porous nanomaterials are still in the laboratory stage. The instrumentation and experimental conditions required for testing experiments are relatively strict. Consequently, combining ECL sensors with

microfluidics and smartphone detection to build portable devices and instruments for environmental detection remains a great challenge.

Author Contributions: C.L. wrote the main text of the manuscript. H.W. and Y.Z. supervise C.L. and gave guidance to write the manuscript. C.L. and J.Y. are responsible for writing the Section 2. R.X., Y.Z. and Q.W. are responsible for writing the Section 3. Y.Z. is responsible for organizing the whole review paper and language editing and wrote the Section 4. Y.Z., R.X. and H.W. provide the fundings for this work. All authors have read and agreed to the published version of the manuscript.

Funding: This work was financially supported by the National Natural Science Foundation of China (21775053), the Natural Science Foundation of Shandong Province (2019GSF111023), and the Applied Basic Research Foundation of Yunnan Province (202201AS070020, 202201AU070061).

Institutional Review Board Statement: Not applicable.

Informed Consent Statement: Not applicable.

Data Availability Statement: Not applicable.

Conflicts of Interest: The authors declare that they have no known competing financial interest or personal relationships that could have appeared to influence the work reported in this paper.

References

1. Chen, S.; Ma, H.; Padelford, J.W.; Qinchen, W.; Yu, W.; Wang, S.; Zhu, M.; Wang, G. Near Infrared Electrochemiluminescence of rod-shape 25-atom AuAg nanoclusters that is hundreds-fold stronger than that of Ru(bpy)$_3^{2+}$ standard. *J. Am. Chem. Soc.* **2019**, *141*, 9603–9609. [CrossRef] [PubMed]
2. Kim, T.; Kim, H.J.; Shin, I.-S.; Hong, J.-I. Potential-dependent electrochemiluminescence for selective molecular sensing of cyanide. *Anal. Chem.* **2020**, *92*, 6019–6025. [CrossRef] [PubMed]
3. Zhang, J.; Jin, R.; Jiang, D.; Chen, H.Y. Electrochemiluminescence-based capacitance microscopy for label-free imaging of antigens on the cellular plasma membrane. *J. Am. Chem. Soc.* **2019**, *141*, 10294–10299. [CrossRef] [PubMed]
4. Hercules David, M. Chemiluminescence resulting from electrochemically generated species. *Science* **1964**, *145*, 808–809. [CrossRef] [PubMed]
5. Santhanam, K.S.V.; Bard, A.J. Chemiluminescence of electrogenerated 9,10-diphenylanthracene anion radical1. *J. Am. Chem. Soc.* **1965**, *87*, 139–140. [CrossRef]
6. Liu, Z.; Qi, W.; Xu, G. Recent advances in electrochemiluminescence. *Chem. Soc. Rev.* **2015**, *44*, 3117–3142. [CrossRef]
7. Miao, W. Electrogenerated chemiluminescence and its biorelated applications. *Chem. Rev.* **2008**, *108*, 2506–2553. [CrossRef] [PubMed]
8. Qi, H.; Zhang, C. Electrogenerated chemiluminescence biosensing. *Anal. Chem.* **2020**, *92*, 524–534. [CrossRef]
9. Chen, Y.; Zhou, S.; Li, L.; Zhu, J.J. Nanomaterials-based sensitive electrochemiluminescence biosensing. *Nano Today* **2017**, *12*, 98–115. [CrossRef]
10. Hu, L.; Xu, G. Applications and trends in electrochemiluminescence. *Chem. Soc. Rev.* **2010**, *39*, 3275–3304. [CrossRef] [PubMed]
11. Ma, C.; Cao, Y.; Gou, X.; Zhu, J.-J. Recent progress in electrochemiluminescence sensing and imaging. *Anal. Chem.* **2020**, *92*, 431–454. [CrossRef]
12. Farka, Z.; Juřík, T.; Kovář, D.; Trnková, L.; Skládal, P. Nanoparticle-based immunochemical biosensors and assays: Recent advances and challenges. *Chem. Rev.* **2017**, *117*, 9973–10042. [CrossRef]
13. Wang, R.; Xi, S.C.; Wang, D.Y.; Dou, M.; Dong, B. Defluorinated porous carbon nanomaterials for CO_2 capture. *ACS Appl. Nano Mater.* **2021**, *4*, 10148–10154. [CrossRef]
14. Jin, L.; Lv, S.; Miao, Y.; Liu, D.; Song, F. Recent development of porous porphyrin-based nanomaterials for photocatalysis. *ChemCatChem* **2021**, *13*, 140–152. [CrossRef]
15. Han, T.; Cao, Y.; Chen, H.Y.; Zhu, J.J. Versatile porous nanomaterials for electrochemiluminescence biosensing: Recent advances and future perspective. *J. Electroanal. Chem.* **2021**, *902*, 115821. [CrossRef]
16. Xu, Y.; Wang, C.; Wu, T.; Ran, G.; Song, Q. Template-free synthesis of porous fluorescent carbon nanomaterials with gluten for intracellular imaging and drug delivery. *ACS Appl. Mater. Interfaces* **2022**, *14*, 21310–21318. [CrossRef]
17. Li, J.; Luo, M.; Yang, H.; Ma, C.; Cai, R.; Tan, W. Novel dual-signal electrochemiluminescence aptasensor involving the resonance energy transform system for kanamycin detection. *Anal. Chem.* **2022**, *94*, 6410–6416. [CrossRef]
18. Wang, S.; Zhao, Y.; Wang, M.; Li, H.; Saqib, M.; Ge, C.; Zhang, X.; Jin, Y. Enhancing luminol electrochemiluminescence by combined use of cobalt-based metal organic frameworks and silver nanoparticles and its application in ultrasensitive detection of cardiac troponin I. *Anal. Chem.* **2019**, *91*, 3048–3054. [CrossRef]
19. Han, Q.; Wang, C.; Liu, P.; Zhang, G.; Song, L.; Fu, Y. Achieving synergistically enhanced dual-mode electrochemiluminescent and electrochemical drug sensors via a multi-effect porphyrin-based metal-organic framework. *Sens. Actuators B Chem.* **2021**, *330*, 129388. [CrossRef]

20. Wang, C.; Liu, L.; Liu, X.; Chen, Y.; Wang, X.; Fan, D.; Kuang, X.; Sun, X.; Wei, Q.; Ju, H. Highly-sensitive electrochemiluminescence biosensor for NT-proBNP using $MoS_2@Cu_2S$ as signal-enhancer and multinary nanocrystals loaded in mesoporous $UiO-66-NH_2$ as novel luminophore. *Sens. Actuators B Chem.* **2020**, *307*, 127619. [CrossRef]
21. Liao, B.Y.; Chang, C.J.; Wang, C.F.; Lu, C.H.; Chen, J.K. Controlled antibody orientation on Fe_3O_4 nanoparticles and CdTe quantum dots enhanced sensitivity of a sandwich-structured electrogenerated chemiluminescence immunosensor for the determination of human serum albumin. *Sens. Actuators B Chem.* **2021**, *336*, 129710. [CrossRef]
22. Wei, X.; Luo, X.; Xu, S.; Xi, F.; Zhao, T. A Flexible Electrochemiluminescence sensor equipped with vertically ordered mesoporous silica nanochannel film for sensitive detection of clindamycin. *Front. Chem.* **2022**, *10*, 2296–2646. [CrossRef] [PubMed]
23. Song, X.; Zhao, L.; Luo, C.; Ren, X.; Yang, L.; Wei, Q. Peptide-based biosensor with a luminescent copper-based metal–organic framework as an electrochemiluminescence emitter for trypsin assay. *Anal. Chem.* **2021**, *93*, 9704–9710. [CrossRef] [PubMed]
24. Li, Y.; Yang, F.; Yuan, R.; Zhong, X.; Zhuo, Y. Electrochemiluminescence covalent organic framework coupling with CRISPR/Cas12a-mediated biosensor for pesticide residue detection. *Food Chem.* **2022**, *389*, 133049. [CrossRef] [PubMed]
25. Ma, Y.; Yu, Y.; Mu, X.; Yu, C.; Zhou, Y.; Chen, J.; Zheng, S.; He, J. Enzyme-induced multicolor colorimetric and electrochemiluminescence sensor with a smartphone for visual and selective detection of Hg^{2+}. *J. Hazard. Mater.* **2021**, *415*, 125538. [CrossRef]
26. Li, X.; Yu, S.; Yan, T.; Zhang, Y.; Du, B.; Wu, D.; Wei, Q. A sensitive electrochemiluminescence immunosensor based on $Ru(bpy)_3^{2+}$ in 3D CuNi oxalate as luminophores and graphene oxide–polyethylenimine as released $Ru(bpy)_3^{2+}$ initiator. *Biosens. Bioelectron.* **2017**, *89*, 1020–1025. [CrossRef]
27. Feng, D.; Wu, Y.; Tan, X.; Chen, Q.; Yan, J.; Liu, M.; Ai, C.; Luo, Y.; Du, F.; Liu, S.; et al. Sensitive detection of melamine by an electrochemiluminescence sensor based on tris(bipyridine)ruthenium(II)-functionalized metal-organic frameworks. *Sens. Actuators B Chem.* **2018**, *265*, 378–386. [CrossRef]
28. Hong, D.; Jo, E.J.; Kim, K.; Song, M.B.; Kim, M.G. $Ru(bpy)_3^{2+}$-Loaded mesoporous silica nanoparticles as electrochemiluminescent probes of a lateral flow immunosensor for highly sensitive and quantitative detection of troponin I. *Small* **2020**, *16*, 2004535. [CrossRef]
29. Zhang, Q.; Liu, Y.; Nie, Y.; Ma, Q.; Zhao, B. Surface plasmon coupling electrochemiluminescence assay based on the use of $AuNP@C_3N_4QD@mSiO_2$ for the determination of the Shiga toxin-producing Escherichia coli (STEC) gene. *Microchim. Acta* **2019**, *186*, 656. [CrossRef]
30. Zhu, H.Y.; Ding, S.N. Dual-signal-amplified electrochemiluminescence biosensor for microRNA detection by coupling cyclic enzyme with CdTe QDs aggregate as luminophor. *Biosens. Bioelectron.* **2019**, *134*, 109–116. [CrossRef]
31. Fang, D.; Zhang, S.; Dai, H.; Lin, Y. An ultrasensitive ratiometric electrochemiluminescence immunosensor combining photothermal amplification for ovarian cancer marker detection. *Biosens. Bioelectron.* **2019**, *146*, 111768. [CrossRef] [PubMed]
32. Li, L.; Zhao, W.; Zhang, J.; Luo, L.; Liu, X.; Li, X.; You, T.; Zhao, C. Label-free Hg(II) electrochemiluminescence sensor based on silica nanoparticles doped with a self-enhanced $Ru(bpy)_3^{2+}$-carbon nitride quantum dot luminophore. *J. Colloid Interface Sci.* **2022**, *608*, 1151–1161. [CrossRef] [PubMed]
33. Xu, C.; Li, J.; Kitte, S.A.; Qi, G.; Li, H.; Jin, Y. Light scattering and luminophore enrichment-enhanced electrochemiluminescence by a 2D porous $Ru@SiO_2$ nanoparticle membrane and its application in ultrasensitive detection of prostate-specific antigen. *Anal. Chem.* **2021**, *93*, 11641–11647. [CrossRef] [PubMed]
34. Pan, D.; Chen, K.; Zhou, Q.; Zhao, J.; Xue, H.; Zhang, Y.; Shen, Y. Engineering of $CdTe/SiO_2$ nanocomposites: Enhanced signal amplification and biocompatibility for electrochemiluminescent immunoassay of alpha-fetoprotein. *Biosens. Bioelectron.* **2019**, *131*, 178–184. [CrossRef]
35. Li, L.; Chen, B.; Luo, L.; Liu, X.; Bi, X.; You, T. Sensitive and selective detection of Hg^{2+} in tap and canal water via self-enhanced ECL aptasensor based on NH_2–$Ru@SiO_2$-NGQDs. *Talanta* **2021**, *222*, 121579. [CrossRef] [PubMed]
36. Li, Z.; Zhang, J.; Chen, H.; Huang, X.; Huang, D.; Luo, F.; Wang, J.; Guo, L.; Qiu, B.; Lin, Z. Electrochemiluminescence biosensor for hyaluronidase based on the $Ru(bpy)_3^{2+}$ doped SiO_2 nanoparticles embedded in the hydrogel fabricated by hyaluronic acid and polyethylenimine. *ACS Appl. Bio Mater.* **2020**, *3*, 1158–1164. [CrossRef] [PubMed]
37. Liang, W.; Zhuo, Y.; Xiong, C.; Zheng, Y.; Chai, Y.; Yuan, R. A sensitive immunosensor via in situ enzymatically generating efficient quencher for electrochemiluminescence of iridium complexes doped SiO_2 nanoparticles. *Biosens. Bioelectron.* **2017**, *94*, 568–574. [CrossRef]
38. Mo, G.; He, X.; Zhou, C.; Ya, D.; Feng, J.; Yu, C.; Deng, B. A novel ECL sensor based on a boronate affinity molecular imprinting technique and functionalized $SiO_2@CQDs/AuNPs/MPBA$ nanocomposites for sensitive determination of alpha-fetoprotein. *Biosens. Bioelectron.* **2019**, *126*, 558–564. [CrossRef]
39. Zhao, W.-R.; Xu, Y.H.; Kang, T.F.; Zhang, X.; Liu, H.; Ming, A.J.; Cheng, S.Y.; Wei, F. Sandwich magnetically imprinted immunosensor for electrochemiluminescence ultrasensing diethylstilbestrol based on enhanced luminescence of $Ru@SiO_2$ by CdTe@ZnS quantum dots. *Biosens. Bioelectron.* **2020**, *155*, 112102. [CrossRef]
40. Li, Y.; Liu, D.; Meng, S.; Zhang, J.; Li, L.; You, T. Regulation of $Ru(bpy)_3^{2+}$ electrochemiluminescence based on distance-dependent electron transfer of ferrocene for dual-signal readout detection of aflatoxin B1 with high sensitivity. *Anal. Chem.* **2022**, *94*, 1294–1301. [CrossRef]
41. Lee, E.J.; Seo, Y.; Park, H.; Kim, M.J.; Yoon, D.; Choung, J.W.; Kim, C.H.; Choi, J.; Lee, K.Y. Development of etched $SiO_2@Pt@ZrO_2$ core-shell catalyst for CO and C_3H_6 oxidation at low temperature. *Appl. Surf. Sci.* **2022**, *575*, 151582. [CrossRef]

42. Gu, Y.; Wu, Y.-n.; Li, L.; Chen, W.; Li, F.; Kitagawa, S. Controllable modular growth of hierarchical MOF-on-MOF architectures. *Angew. Chem. Int. Ed.* **2017**, *56*, 15658–15662. [CrossRef] [PubMed]
43. Long, J.; Gong, Y.; Lin, J. Metal–organic framework–derived Co_9S_8@CoS@CoO@C nanoparticles as efficient electro- and photocatalysts for the oxygen evolution reaction. *J. Mater. Chem. A* **2017**, *5*, 10495–10509. [CrossRef]
44. Fonseca, J.; Gong, T.; Jiao, L.; Jiang, H.-L. Metal–organic frameworks (MOFs) beyond crystallinity: Amorphous MOFs, MOF liquids and MOF glasses. *J. Mater. Chem. A* **2021**, *9*, 10562–10611. [CrossRef]
45. Miao, Y.B.; Ren, H.X.; Zhong, Q.; Song, F.X. Tailoring a luminescent metal−organic framework precise inclusion of Pt-Aptamer nanoparticle for noninvasive monitoring Parkinson's disease. *Chem. Eng. J.* **2022**, *441*, 136009. [CrossRef]
46. Wang, C.; Zhang, N.; Wei, D.; Feng, R.; Fan, D.; Hu, L.; Wei, Q.; Ju, H. Double electrochemiluminescence quenching effects of Fe_3O_4@PDA-Cu_XO towards self-enhanced $Ru(bpy)_3^{2+}$ functionalized MOFs with hollow structure and it application to procalcitonin immunosensing. *Biosens. Bioelectron.* **2019**, *142*, 111521. [CrossRef]
47. Yang, X.; Yu, Y.Q.; Peng, L.Z.; Lei, Y.M.; Chai, Y.Q.; Yuan, R.; Zhuo, Y. Strong electrochemiluminescence from MOF accelerator enriched quantum dots for enhanced sensing of trace cTnI. *Anal. Chem.* **2018**, *90*, 3995–4002. [CrossRef]
48. Dong, X.; Zhao, G.; Li, X.; Fang, J.; Miao, J.; Wei, Q.; Cao, W. Electrochemiluminescence immunosensor of "signal-off" for β-amyloid detection based on dual metal-organic frameworks. *Talanta* **2020**, *208*, 120376. [CrossRef]
49. Mo, G.; Qin, D.; Jiang, X.; Zheng, X.; Mo, W.; Deng, B. A sensitive electrochemiluminescence biosensor based on metal-organic framework and imprinted polymer for squamous cell carcinoma antigen detection. *Sens. Actuators B Chem.* **2020**, *310*, 127852. [CrossRef]
50. Dong, X.; Zhao, G.; Liu, L.; Li, X.; Wei, Q.; Cao, W. Ultrasensitive competitive method-based electrochemiluminescence immunosensor for diethylstilbestrol detection based on $Ru(bpy)_3^{2+}$ as luminophor encapsulated in metal–organic frameworks UiO-67. *Biosens. Bioelectron.* **2018**, *110*, 201–206. [CrossRef]
51. Nie, Y.; Tao, X.; Zhang, H.; Chai, Y.Q.; Yuan, R. Self-assembly of gold nanoclusters into a metal–organic framework with efficient electrochemiluminescence and their application for sensitive detection of rutin. *Anal. Chem.* **2021**, *93*, 3445–3451. [CrossRef] [PubMed]
52. Zhao, Q.; Wang, Y.; Li, X.; Yue, Q.; Dong, X.; Du, B.; Cao, W.; Wei, Q. Dual-quenching electrochemiluminescence strategy based on three-dimensional metal-organic frameworks for ultrasensitive detection of amyloid-β. *Anal. Chem.* **2019**, *91*, 1989–1996. [CrossRef] [PubMed]
53. Dong, X.; Du, Y.; Zhao, G.; Cao, W.; Fan, D.; Kuang, X.; Wei, Q.; Ju, H. Dual-signal electrochemiluminescence immunosensor for Neuron-specific enolase detection based on "dual-potential" emitter $Ru(bpy)_3^{2+}$ functionalized zinc-based metal-organic frameworks. *Biosens. Bioelectron.* **2021**, *192*, 113505. [CrossRef]
54. Liu, H.; Liu, Z.; Yi, J.; Ma, D.; Xia, F.; Tian, D.; Zhou, C. A dual-signal electroluminescence aptasensor based on hollow Cu/Co-MOF-luminol and g-C_3N_4 for simultaneous detection of acetamiprid and malathion. *Sens. Actuators B Chem.* **2021**, *331*, 129412. [CrossRef]
55. Huang, W.; Hu, G.B.; Liang, W.B.; Wang, J.M.; Lu, M.L.; Yuan, R.; Xiao, D.R. Ruthenium(II) complex-grafted hollow hierarchical metal–organic frameworks with superior electrochemiluminescence performance for sensitive assay of thrombin. *Anal. Chem.* **2021**, *93*, 6239–6245. [CrossRef]
56. Wen, J.; Jiang, D.; Shan, X.; Wang, W.; Xu, F.; Shiigi, H.; Chen, Z. Ternary electrochemiluminescence biosensor based on black phosphorus quantum dots doped perylene derivative and metal organic frameworks as a coreaction accelerator for the detection of chloramphenicol. *Microchem. J.* **2022**, *172*, 106927. [CrossRef]
57. Xia, M.; Zhou, F.; Feng, X.; Sun, J.; Wang, L.; Li, N.; Wang, X.; Wang, G. A DNAzyme-based dual-stimuli responsive electrochemiluminescence resonance energy transfer platform for ultrasensitive anatoxin-a detection. *Anal. Chem.* **2021**, *93*, 11284–11290. [CrossRef]
58. Fang, J.; Li, J.; Feng, R.; Yang, L.; Zhao, L.; Zhang, N.; Zhao, G.; Yue, Q.; Wei, Q.; Cao, W. Dual-quenching electrochemiluminescence system based on novel acceptor CoOOH@Au NPs for early detection of procalcitonin. *Sens. Actuators B Chem.* **2021**, *332*, 129544. [CrossRef]
59. Huang, L.Y.; Hu, X.; Shan, H.-Y.; Yu, L.; Gu, Y.X.; Wang, A.J.; Shan, D.; Yuan, P.X.; Feng, J.J. High-performance electrochemiluminescence emitter of metal organic framework linked with porphyrin and its application for ultrasensitive detection of biomarker mucin-1. *Sens. Actuators B Chem.* **2021**, *344*, 130300. [CrossRef]
60. Hu, G.B.; Xiong, C.Y.; Liang, W.B.; Zeng, X.S.; Xu, H.L.; Yang, Y.; Yao, L.Y.; Yuan, R.; Xiao, D.R. Highly stable mesoporous luminescence-functionalized mof with excellent electrochemiluminescence property for ultrasensitive immunosensor construction. *ACS Appl. Mater. Interfaces* **2018**, *10*, 15913–15919. [CrossRef]
61. Wang, Y.; Zhao, G.; Chi, Y.; Yang, S.; Niu, Q.; Wu, D.; Cao, W.; Li, T.; Ma, H.; Wei, Q. Self-luminescent lanthanide metal–organic frameworks as signal probes in electrochemiluminescence immunoassay. *J. Am. Chem. Soc.* **2021**, *143*, 504–512. [CrossRef] [PubMed]
62. Yan, M.; Ye, J.; Zhu, Q.; Zhu, L.; Huang, J.; Yang, X. Ultrasensitive immunosensor for cardiac troponin i detection based on the electrochemiluminescence of 2D Ru-MOF nanosheets. *Anal. Chem.* **2019**, *91*, 10156–10163. [CrossRef] [PubMed]
63. Zhao, L.; Wang, M.; Song, X.; Liu, X.; Ju, H.; Ai, H.; Wei, Q.; Wu, D. Annihilation luminescent Eu-MOF as a near-infrared electrochemiluminescence probe for trace detection of trenbolone. *Chem. Eng. J.* **2022**, *434*, 134691. [CrossRef]

64. Li, J.; Jia, H.; Ren, X.; Li, Y.; Liu, L.; Feng, R.; Ma, H.; Wei, Q. Dumbbell plate-shaped AIEgen-based luminescent MOF with high quantum yield as self-enhanced ECL tags: Mechanism insights and biosensing application. *Small* **2022**, *18*, 2106567. [CrossRef]
65. Huang, W.; Hu, G.B.; Yao, L.Y.; Yang, Y.; Liang, W.B.; Yuan, R.; Xiao, D.R. Matrix coordination-induced electrochemiluminescence enhancement of tetraphenylethylene-based hafnium metal–organic framework: An electrochemiluminescence chromophore for ultrasensitive electrochemiluminescence sensor construction. *Anal. Chem.* **2020**, *92*, 3380–3387. [CrossRef] [PubMed]
66. Yao, L.Y.; Yang, F.; Hu, G.B.; Yang, Y.; Huang, W.; Liang, W.B.; Yuan, R.; Xiao, D.R. Restriction of intramolecular motions (RIM) by metal-organic frameworks for electrochemiluminescence enhancement:2D Zr_{12}-adb nanoplate as a novel ECL tag for the construction of biosensing platform. *Biosens. Bioelectron.* **2020**, *155*, 112099. [CrossRef]
67. Zhou, L.; Yang, L.; Wang, C.; Jia, H.; Xue, J.; Wei, Q.; Ju, H. Copper doped terbium metal organic framework as emitter for sensitive electrochemiluminescence detection of CYFRA 21-1. *Talanta* **2022**, *238*, 123047. [CrossRef]
68. Wang, J.M.; Yao, L.Y.; Huang, W.; Yang, Y.; Liang, W.B.; Yuan, R.; Xiao, D.R. Overcoming aggregation-induced quenching by metal−organic framework for electrochemiluminescence (ECL) enhancement: Zn-PTC as a new ECL emitter for ultrasensitive MicroRNAs detection. *ACS Appl. Mater. Interfaces* **2021**, *13*, 44079–44085. [CrossRef]
69. Yao, L.Y.; Yang, F.; Liang, W.B.; Hu, G.B.; Yang, Y.; Huang, W.; Yuan, R.; Xiao, D.R. Ruthenium complex doped metal-organic nanoplate with high electrochemiluminescent intensity and stability for ultrasensitive assay of mucin 1. *Sens. Actuators B Chem.* **2019**, *292*, 105–110. [CrossRef]
70. Wang, C.; Li, Z.; Ju, H. Copper-doped terbium luminescent metal organic framework as an emitter and a co-reaction promoter for amplified electrochemiluminescence immunoassay. *Anal. Chem.* **2021**, *93*, 14878–14884. [CrossRef] [PubMed]
71. Ma, X.; Pang, C.; Li, S.; Li, J.; Wang, M.; Xiong, Y.; Su, L.; Luo, J.; Xu, Z.; Lin, L. Biomimetic synthesis of ultrafine mixed-valence metal–organic framework nanowires and their application in electrochemiluminescence sensing. *ACS Appl. Mater. Interfaces* **2021**, *13*, 41987–41996. [CrossRef]
72. Li, J.; Luo, M.; Jin, C.; Zhang, P.; Yang, H.; Cai, R.; Tan, W. Plasmon-enhanced electrochemiluminescence of PTP-decorated Eu MOF-Based Pt-tipped Au bimetallic nanorods for the lincomycin assay. *ACS Appl. Mater. Interfaces* **2022**, *14*, 383–389. [CrossRef] [PubMed]
73. Zhou, Y.; He, J.; Zhang, C.; Li, J.; Fu, X.; Mao, W.; Li, W.; Yu, C. Novel Ce(III)-metal organic framework with a luminescent property to fabricate an electrochemiluminescence immunosensor. *ACS Appl. Mater. Interfaces* **2020**, *12*, 338–346. [CrossRef] [PubMed]
74. Wang, X.; Xiao, S.; Yang, C.; Hu, C.; Wang, X.; Zhen, S.; Huang, C.; Li, Y. Zinc–metal organic frameworks: A coreactant-free electrochemiluminescence luminophore for ratiometric detection of miRNA-133a. *Anal. Chem.* **2021**, *93*, 14178–14186. [CrossRef] [PubMed]
75. Liu, Q.; Yang, Y.; Liu, X.P.; Wei, Y.P.; Mao, C.J.; Chen, J.S.; Niu, H.L.; Song, J.M.; Zhang, S.Y.; Jin, B.K.; et al. A facile in situ synthesis of MIL-101-CdSe nanocomposites for ultrasensitive electrochemiluminescence detection of carcinoembryonic antigen. *Sens. Actuators B Chem.* **2017**, *242*, 1073–1078. [CrossRef]
76. Sun, M.F.; Liu, J.L.; Zhou, Y.; Zhang, J.Q.; Chai, Y.Q.; Li, Z.H.; Yuan, R. High-efficient electrochemiluminescence of BCNO quantum dot-equipped boron active sites with unexpected catalysis for ultrasensitive detection of MicroRNA. *Anal. Chem.* **2020**, *92*, 14723–147299. [CrossRef] [PubMed]
77. Zhang, Q.; Liu, Y.; Nie, Y.; Liu, Y.; Ma, Q. Wavelength-dependent surface plasmon coupling electrochemiluminescence biosensor based on sulfur-doped carbon nitride quantum dots for K-RAS gene detection. *Anal. Chem.* **2019**, *91*, 13780–13786. [CrossRef]
78. Feng, D.; Wei, F.; Wu, Y.; Tan, X.; Li, F.; Lu, Y.; Fan, G.; Han, H. A novel signal amplified electrochemiluminescence biosensor based on MIL-53(Al)@CdS QDs and SiO_2@AuNPs for trichlorfon detection. *Analyst* **2021**, *146*, 1295–1302. [CrossRef]
79. Shan, X.; Pan, T.; Pan, Y.; Wang, W.; Chen, X.; Shan, X.; Chen, Z. Highly sensitive and selective detection of Pb(II) by $NH_2-SiO_2/Ru(bpy)_3^{2+}$–UiO66 based solid-state ECL sensor. *Electroanalysis* **2020**, *32*, 462–469. [CrossRef]
80. Cai, M.; Loague, Q.R.; Zhu, J.; Lin, S.; Usov, P.M.; Morris, A.J. Ruthenium(ii)-polypyridyl doped zirconium(iv) metal–organic frameworks for solid-state electrochemiluminescence. *Dalton Trans.* **2018**, *47*, 16807–16812. [CrossRef] [PubMed]
81. Zhao, L.; Song, X.; Ren, X.; Wang, H.; Fan, D.; Wu, D.; Wei, Q. Ultrasensitive near-infrared electrochemiluminescence biosensor derived from Eu-MOF with antenna effect and high efficiency catalysis of specific CoS_2 hollow triple shelled nanoboxes for procalcitonin. *Biosens. Bioelectron.* **2021**, *191*, 113409. [CrossRef]
82. Yang, Z.R.; Wang, M.M.; Wang, X.S.; Yin, X.B. Boric-acid-functional lanthanide metal–organic frameworks for selective ratiometric fluorescence detection of fluoride ions. *Anal. Chem.* **2017**, *89*, 1930–1936. [CrossRef] [PubMed]
83. Zhu, D.; Zhang, Y.; Bao, S.; Wang, N.; Yu, S.; Luo, R.; Ma, J.; Ju, H.; Lei, J. Dual intrareticular oxidation of mixed-ligand metal–organic frameworks for stepwise electrochemiluminescence. *J. Am. Chem. Soc.* **2021**, *143*, 3049–3053. [CrossRef]
84. Carrara, S.; Aliprandi, A.; Hogan, C.F.; De Cola, L. Aggregation-induced electrochemiluminescence of platinum(II) complexes. *J. Am. Chem. Soc.* **2017**, *139*, 14605–14610. [CrossRef] [PubMed]
85. Yang, Y.; Hu, G.B.; Liang, W.B.; Yao, L.Y.; Huang, W.; Zhang, Y.J.; Zhang, J.L.; Wang, J.M.; Yuan, R.; Xiao, D.R. An AIEgen-based 2D ultrathin metal–organic layer as an electrochemiluminescence platform for ultrasensitive biosensing of carcinoembryonic antigen. *Nanoscale* **2020**, *12*, 5932–5941. [CrossRef] [PubMed]
86. Li, J.; Jing, X.; Li, Q.; Li, S.; Gao, X.; Feng, X.; Wang, B. Bulk COFs and COF nanosheets for electrochemical energy storage and conversion. *Chem. Soc. Rev.* **2020**, *49*, 3565–3604. [CrossRef] [PubMed]

87. Zeng, W.J.; Wang, K.; Liang, W.B.; Chai, Y.Q.; Yuan, R.; Zhuo, Y. Covalent organic frameworks as micro-reactors: Confinement-enhanced electrochemiluminescence. *Chem. Sci.* **2020**, *11*, 5410–5414. [CrossRef] [PubMed]
88. Zhang, J.L.; Yao, L.Y.; Yang, Y.; Liang, W.B.; Yuan, R.; Xiao, D.R. Conductive covalent organic frameworks with conductivity and pre-reduction-enhanced electrochemiluminescence for ultrasensitive biosensor construction. *Anal. Chem.* **2022**, *94*, 3685–3692. [CrossRef] [PubMed]
89. Ravikumar, A.; Panneerselvam, P. A novel fluorescent sensing platform based on metal-polydopamine frameworks for the dual detection of kanamycin and oxytetracycline. *Analyst* **2019**, *144*, 2337–2344.
90. Ren, R.; Cai, G.; Yu, Z.; Zeng, Y.; Tang, D. Metal-polydopamine framework: An innovative signal-generation tag for colorimetric immunoassay. *Anal. Chem.* **2018**, *90*, 11099–11105. [CrossRef] [PubMed]
91. Ravikumar, A.; Panneerselvam, P.; Morad, N. Metal–polydopamine framework as an effective fluorescent quencher for highly sensitive detection of Hg(II) and Ag(I) ions through exonuclease III activity. *ACS Appl. Mater. Interfaces* **2018**, *10*, 20550–20558. [CrossRef] [PubMed]
92. Chen, A.; Zhao, M.; Zhuo, Y.; Chai, Y.; Yuan, R. Hollow porous polymeric nanospheres of a self-enhanced ruthenium complex with improved electrochemiluminescent efficiency for ultrasensitive aptasensor construction. *Anal. Chem.* **2017**, *89*, 9232–9238. [CrossRef] [PubMed]
93. Zhao, B.; Luo, Y.; Qu, X.; Hu, Q.; Zou, J.; He, Y.; Liu, Z.; Zhang, Y.; Bao, Y.; Wang, W.; et al. Graphite-like carbon nitride nanotube for electrochemiluminescence featuring high efficiency, high stability, and ultrasensitive ion detection capability. *J. Phys. Chem. Lett.* **2021**, *12*, 11191–11198. [CrossRef] [PubMed]
94. Malik, L.A.; Bashir, A.; Qureashi, A.; Pandith, A.H. Detection and removal of heavy metal ions: A review. *Environ. Chem. Lett.* **2019**, *17*, 1495–1521. [CrossRef]
95. Liao, H.; Jin, C.; Zhou, Y.; Chai, Y.; Yuan, R. Novel ABEI/dissolved O_2/Ag_3BiO_3 nanocrystals ECL ternary system with high luminous efficiency for ultrasensitive determination of MicroRNA. *Anal. Chem.* **2019**, *91*, 11447–11454. [CrossRef] [PubMed]
96. Gao, T.B.; Zhang, J.J.; Wen, J.; Yang, X.X.; Ma, H.B.; Cao, D.K.; Jiang, D. Single-molecule MicroRNA electrochemiluminescence detection using cyclometalated dinuclear Ir(III) complex with synergistic effect. *Anal. Chem.* **2020**, *92*, 1268–1275. [CrossRef]
97. Zhang, J.L.; Yang, Y.; Liang, W.B.; Yao, L.Y.; Yuan, R.; Xiao, D.R. Highly stable covalent organic framework nanosheets as a new generation of electrochemiluminescence emitters for ultrasensitive MicroRNA detection. *Anal. Chem.* **2021**, *93*, 3258–3265. [CrossRef] [PubMed]
98. Jiang, Y.; Li, R.; He, W.; Li, Q.; Yang, X.; Li, S.; Bai, W.; Li, Y. MicroRNA-21 electrochemiluminescence biosensor based on Co-MOF–N-(4-aminobutyl)-N-ethylisoluminol/$Ti_3C_2T_x$ composite and duplex-specific nuclease-assisted signal amplification. *Microchim. Acta* **2022**, *189*, 129. [CrossRef]
99. Jian, Y.; Wang, H.; Lan, F.; Liang, L.; Ren, N.; Liu, H.; Ge, S.; Yu, J. Electrochemiluminescence based detection of microRNA by applying an amplification strategy and Hg(II)-triggered disassembly of a metal organic frameworks functionalized with ruthenium(II)tris(bipyridine). *Microchim. Acta* **2018**, *185*, 133. [CrossRef] [PubMed]

Article

Label-Free and Homogeneous Electrochemical Biosensor for Flap Endonuclease 1 Based on the Target-Triggered Difference in Electrostatic Interaction between Molecular Indicators and Electrode Surface

Jianping Zheng [1], Xiaolin Xu [2], Hanning Zhu [2], Zhipeng Pan [2], Xianghui Li [2,*], Fang Luo [3] and Zhenyu Lin [3,*]

[1] Department of Oncology, Shengli Clinical Medical College, Fujian Medical University, Fujian Provincial Hospital, Fuzhou 350001, China; fjslzjp@sina.com
[2] Department of Clinical Laboratory, School of Medical Technology and Engineering, Fujian Medical University, Fuzhou 350004, China; 1696219821@fjmu.edu.cn (X.X.); zhuhanning95@163.com (H.Z.); panzp0109@163.com (Z.P.)
[3] Ministry of Education Key Laboratory for Analysis Science of Food Safety and Biology, Fujian Provincial Key Laboratory of Analysis and Detection for Food Safety, College of Chemistry, Fuzhou University, Fuzhou 350116, China; luofang@fzu.edu.cn
* Correspondence: lixianghui1987@126.com (X.L.); zylin@fzu.edu.cn (Z.L.); Tel./Fax: +86-591-22866135 (X.L. & Z.L.)

Abstract: Target-induced differences in the electrostatic interactions between methylene blue (MB) and indium tin oxide (ITO) electrode surface was firstly employed to develop a homogeneous electrochemical biosensor for flap endonuclease 1 (FEN1) detection. In the absence of FEN1, the positively charged methylene blue (MB) is free in the solution and can diffuse onto the negatively charged ITO electrode surface easily, resulting in an obvious electrochemical signal. Conversely, with the presence of FEN1, a 5′-flap is cleaved from the well-designed flapped dumbbell DNA probe (FDP). The remained DNA fragment forms a closed dumbbell DNA probe to trigger hyperbranched rolling circle amplification (HRCA) reaction, generating plentiful dsDNA sequences. A large amount of MB could be inserted into the produced dsDNA sequences to form MB-dsDNA complexes, which contain a large number of negative charges. Due to the strong electrostatic repulsion between MB-dsDNA complexes and the ITO electrode surface, a significant signal drop occurs. The signal change ($\Delta Current$) shows a linear relationship with the logarithm of FEN1 concentration from 0.04 to 80.0 U/L with a low detection limit of 0.003 U/L (S/N = 3). This study provides a label-free and homogeneous electrochemical platform for evaluating FEN1 activity.

Keywords: homogeneous; electrochemical biosensor; label-free; hyperbranched rolling circle amplification; flap endonuclease 1

1. Introduction

Flap endonuclease 1 (FEN1) exhibits multiple values in the early diagnosis [1,2], targeting therapy [3–6], and prognostic monitoring [7,8] of various cancers. Traditional assays, including western blot, reverse transcription-polymerase chain reaction (RT-PCR), and enzyme-linked immunosorbent assay (ELISA) [1,3,4] had already been utilized to detect FEN1. Several novel strategies were also designed for FEN1 detection [9–15]. For instance, Zhang et al. [9] proposed a DNA-based fluorescent biosensor to evaluate FEN1 activity in living cells. Our group developed an electrochemiluminescence (ECL) biosensor for FEN1 via combining branched hybridization chain reaction (BHCR) amplification, ultrafiltration separation, and ECL detection [16]. Although these methods can realize FEN1 detection with high accuracy, it is still desirable to explore novel analytical methods with increased performance and decreased cost.

Electrochemical analysis combining the merits of low cost, simple operation, high sensitivity, and fast response has been extensively used in biological and biomedical applications [17,18]. However, most electrochemical sensors require the laborious and time-consuming immobilization of recognition probes on the electrode surface, labeling of reporter molecules first. Additionally, target recognition takes place on the interface between solution and electrode, which lowers the reaction efficiency and recognition rates because of the steric hindrance. Immobilization-free homogeneous electrochemical methods have been developed on the basis of the difference in the electrostatic interaction between DNA sequences and the negatively charged indium tin oxide (ITO) electrode can address these concerns well [19–21]. Methylene Blue (MB), as a widely used electrochemical indicator, is a positively charged organic dye that can be inserted into double-stranded DNA (dsDNA) through π–π stacking interactions [22]. Taking advantage of its special interaction with dsDNA, MB has been adopted to design an enzyme-free and label-free homogeneous electrochemical miRNA biosensor via the difference in electrostatic repulsion between MB-intercalated dsDNA and ITO electrode [23]. However, as far as we know, this strategy had not been applied to detect FEN1 activity.

Hyperbranched rolling circle amplification (HRCA) [24] evolved from rolling circle amplification (RCA) and is a simple and convenient method with much higher isothermal amplification efficiency (10^9-fold) than that of RCA [25]. This can produce a large number of dsDNAs with high efficiency. In this study, a label-free and homogeneous electrochemical biosensor has been designed for monitoring FEN1 activity by combining the well-designed flapped dumbbell DNA probe (FDP), target-induced electrostatic interactions between MB molecules, and negatively charged dsDNA strands, and the excellent HRCA technology. This homogeneous biosensor can avoid the complicated electrode modification, high cost of labeling, and steric hindrance during the HRCA, which guarantees that the detection process will be simpler and faster, while achieving low costs and good reproducibility. Additionally, the proposed biosensor was used to measure the FEN1 levels in clinical samples with satisfied performance, and therefore, could serve as a potent platform for monitoring FEN1 activity in clinical diagnosis.

2. Experimental Section

2.1. Reagents and Oligonucleotides

The lysates of AGS and HaCaT cells were prepared and stored at $-20\ ^{\circ}\text{C}$ for the following assays (see the details in the Supplementary Information (SI)). All the other reagents and chemicals were analytical grade. Ultrapure water with a resistance of 18.2 MΩ·cm was adopted in the whole experiment. The designed oligonucleotides were synthesized by Shanghai Sangon Biotechnology Co., Ltd. (Shanghai, China) with the following sequences:

FDP: 5′-TTAACGACCATTCAAACGCACTGATGGTTGCCAACCACAAACGGCA A-3′
P1: 5′-CAGTGCGTT-3′
P2: 5′-ACCACAAAC-3′

The other materials and reagents employed in this experiment are listed in the SI.

2.2. Apparatus

Differential pulse voltammetry (DPV) signals were measured by a CHI660a electrochemical workstation (Chenhua Instruments, Shanghai, China). ITO electrode served as a working electrode, and platinum wires were utilized as the reference electrode and the auxiliary electrode, respectively. ITO electrodes were bought from Huanan Xiangcheng Technology Co., Ltd. (Shenzhen, China). Before DPV signals were acquired, the surface of the ITO electrode was firstly decorated with negative charges by sequentially sonicating it in Alconox solution (20 g/L), 2-propanol, and ultrapure water for 15 min. The ITO electrode was then inserted vertically into the Nafion solution (0.5 mg/mL) for 10 s and then removed and dried to quickly prepare a negatively charged electrode. The operating area of the ITO electrode was set to 3 mm × 3 mm.

In total, 12% polyacrylamide gel electrophoresis was employed to notarize the occurrence of each course in the HRCA reaction. Gel electrophoresis imager (JS-2012) was obtained from Shanghai Peiqing Technology Co., Ltd. (Shanghai, China). The test was performed in 0.5× triborate-EDTA (TBE) buffer (pH 8.4) under a constant voltage of 90 V for 60 min at room temperature. SYBR Green I (20×) was added to the amplification product and incubated at room temperature in the dark for 15 min. Then, the fluorescence signals were detected by an F-4600 Fluorescence spectrofluorometer purchased from Hitachi high-technologies corporation (Tokyo, Japan) in the range of 500~650 nm with the excitation wavelength setting to 488 nm.

2.3. Process of FEN1 Detection

The test sample (10 µL), FDP (10 nM), and reaction buffer (3 mM $MgSO_4$, 15 mM $(NH_4)_2SO_4$, 30 mM Tris-HCl, and 15 mM KCl, pH 8.8) of 30 µL were mixed and incubated at 37 °C for 70 min. Afterward, T4 DNA ligase (2 µL, 5 U/µL) was added to the system and incubated at 22 °C for 70 min with a corresponding 1× reaction buffer. Then, the superfluous ssDNA and dsDNA probes were removed by Exo I (10 U) and Exo III (20 U) at 37 °C for 1 h and further inactivated at 80 °C for 15 min. Subsequently, primers (P1, P2, 1 µM), dNTPs (0.4 mM), Bst DNA polymerase (80 U/mL), and 1× reaction buffer were mixed and amplified for 100 min at 63 °C. Finally, MB (40 µM) was added to the 200 µL total reaction system, and the current signal was detected.

3. Results and Discussion

3.1. Principle of the Proposed Biosensor for FEN1

Scheme 1 clearly exhibits the principle of the designed strategy for FEN1 detection. Firstly, FDP with a 5′-flap that can be identified and cleaved by FEN1 was rationally designed. Free MB molecules have positive charges, and the ITO electrode carries negative charges. Thus, MB can diffuse freely onto ITO to generate high electrochemical signals. In the presence of FEN1, 5′-flap is separated from FDP and leaves an exposed 5′ phosphate group. Due to the stabilization of DNA scaffold the 5′ phosphate approaches the 3′-flap to form a nick site that is ligatable by T4 DNA ligase. After ligation, a closed dumbbell DNA probe (C-DNA) with a circular conformation is formed. Then, only C-DNA probes remained by adding Exo I and III to the solution. Afterward, the HRCA mixture, including primers, dNTPs, Bst DNA polymerase, and MB, was added to the above solution. After initiating the HRCA reaction, a large amount of dsDNA was produced, and abundant MB can be inserted into these dsDNA products to form MB-dsDNA complexes. The resulting MB-dsDNA complexes carry negative charges. Hence, a significantly reduced electrochemical signal was detected. Therefore, the recorded electrochemical signal of this system is related to the concentration of the FEN1 target, resulting in a label-free and immobilization-free electrochemical biosensor for detecting the FEN1 activity. In contrast, without FEN1, the 5′-flap could not be cleaved and the ligation would not occur. So, the current signal of the solution hardly changed.

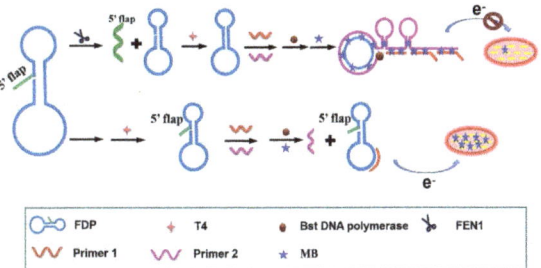

Scheme 1. Scheme of the homogeneous electrochemical biosensor for FEN1 detection based on the difference in electrostatic interaction.

3.2. Feasibility Test

First of all, polyacrylamide gel electrophoresis, as the gold standard to analyze nucleic acids for length, bend, and flexibility was used to ensure the feasibility of the HRCA strategy. As displayed in Figure 1A, seven lanes were observed in the image of polyacrylamide gel electrophoresis. Lane a with a single bright band represents FDP in the monomeric and uniform state. Upon the addition of FEN1, the 5′-flap of FDP can be cleaved by the FEN1 target, forming a dumbbell DNA and a short DNA. Thus, lane b contains two obvious bands, in which the lower band corresponds to the dissociated flap sequences and the upper band denotes the closed dumbbell-shaped DNA. This lane clearly evidences the specific recognition ability of FEN1 to cleave the 5′-flap structure on FDP. Similar to lane b, the mixture of FDP, FEN1, T4 DNA ligase, and the necessary buffer also generated two bands in lane c. Different from lane c, only one bright band is left in lane d after adding Exo I and III, with the same position as the uppermost band in lane c. This verifies the above conjecture that a closed circular dumbbell-shaped DNA is indeed generated under the action of T4 ligase. Finally, with the addition of the HRCA mixture, a band with tailing appears in lane f, which corresponds to the products of the HRCA reaction. Conversely, without FEN1 targets, no 5′-flap can be cleaved from FDP, and thus the HRCA cannot be induced to occur. Therefore, no obvious bands emerged in lane e. All in all, these results clearly revealed that FEN1 indeed could cleave the 5′-flap of FDP to initiate the subsequent HRCA process.

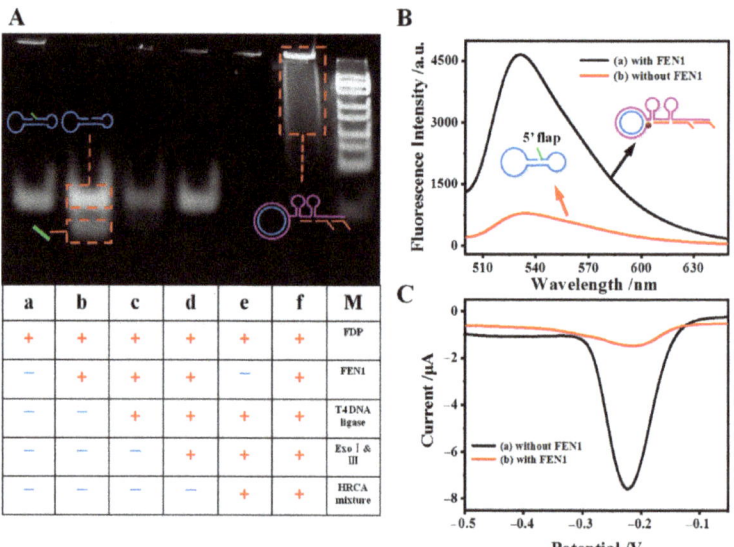

Figure 1. Feasibility study. (**A**) Polyacrylamide gel electrophoresis under different conditions: (a) FDP, (b) FDP + FEN1, (c) FDP + FEN1 + T4, (d) FDP + FEN1 + T4 + Exo I and III, (e) FDP + T4 + Exo I and III + HRCA mixture, (f) FDP + FEN1 + T4 + Exo I and III + HRCA mixture, (M) marker. (**B**) Fluorescence spectra at present and absence of FEN1: (a) with FEN1, (b) without FEN1. (**C**) The DPV responses with and without target: a) without FEN1; b) with FEN1. The concentrations of FEN1 and FDP are 80 U/L and 5 nM, respectively.

In addition, fluorescence spectra have been also utilized to verify the proposed sensing scheme (Figure 1B). SYBR Green I could be embedded into the holes of dsDNA, yielding a strong fluorescent intensity at 530 nm. When FEN1 is added, an obvious fluorescence emission could be recorded at 530 nm (curve a), indicating that plentiful dsDNAs were produced. In contrast, in the absence of FEN1, very low fluorescence intensity was recorded (curve b) because the FDP also has a certain amount of complementary structure. It is demonstrated that the FDP, T4 DNA ligase, and HRCA mixture are not capable to induce HRCA to generate a large amount of dsDNA.

The feasibility of this sensor can be also verified by the change in the zeta potential of the studied system before and after the reaction. As shown in Figure S1, a negative potential of −40.99 mV was observed for the product of the sample with FEN1, which can be attributed to the negatively charged phosphate on the amount of dsDNA molecular skeleton. Concurrently, a negative potential of −8.06 mV was obtained for the product of the sample without FEN1, indicating that no amplification reaction occurred in this system. After incubating with positively charged MB, the zeta potential of the experimental group changed to −26.65 mV, while the zeta potential of the control group turned into 2.62 mV. This fact confirms the completion of the amplification reaction and successful embedding of MB.

Furthermore, DPV responses were tested to examine this proposed sensing scheme. As shown in Figure 1C, the DPV signal is significant in the absence of FEN1 (curve a). This is because the amplification reaction cannot be initiated, which will not affect the MB in the solution to freely diffuse to the electrode surface. When the FEN1 targets are present, an obvious decrease in DPV signal can be found (curve b) because the HRCA reaction generates a great amount of dsDNA and the DPV signal reporter MB molecules are mainly embedded into the resulting dsDNA sequences. Thus, only a fraction of remaining MB molecules can diffuse onto the ITO electrode surface to yield a much weaker DPV signal.

3.3. Optimization of the Experimental Conditions

To reveal the best performance of this homogeneous electrochemical sensor, the influence of FEN1 digestion time, HRCA reaction, and the concentration of MB on the final readout have been investigated. Amongst these, the concentration of MB is a vital factor for the performance of biosensors because an appropriate concentration of MB could assure a satisfactory detection range and low background response. Firstly, the HRCA products were treated with different concentrations of MB for 1 h. From Figure 2A, it can be seen that the DPV responses of both the blank sample and the one with FEN1 of 0.8 U/L gradually increase with the increasing MB concentration. However, $\Delta Current$ values, namely the difference in current in absence and presence of FEN1, progressively increase and subsequently decrease slightly when the concentration of MB reaches 40 µM. Thereby, the optimal MB concentration is settled as 40 µM.

The dosage of Bst DNA polymerase can directly affect the efficiency of HRCA. Figure 2B shows that, along with the raising dosage of Bst DNA polymerase, the DPV peak current sharply declines until 80 U/mL. After that, the DPV signals hardly change anymore, which reveals that 80 U/mL of Bst DNA polymerase is suitable for the HRCA reaction. For HRCA reaction, the concentration of dNTPs is also pivotal. As shown in Figure 2C, when the number of used dNTPs is more than 0.4 mM, the DPV current of the sample stops decreasing, indicating that 0.4 mM is the minimum amount of dNTPs needed for the HRCA reaction.

Subsequently, different digestion time intervals and reaction times were evaluated. As presented in Figure 2D, the current response firstly drops, and then reaches a plateau after 70 min of digestion, suggesting that 70 min is long enough for completely cleaving the flap of FDP. It is also observed from Figure 2D that, before reaching the lowest point at 100 min, the current is almost proportional to the HRCA reaction time. The current hardly changes after 100 min of the HRCA reaction, which implied that 100 min is enough for the amplification reaction. Such tests clearly demonstrate that the best digestion duration is 70 min, and the suitable HRCA reaction time is 100 min. Hence, these optimized conditions have been used in the following assays.

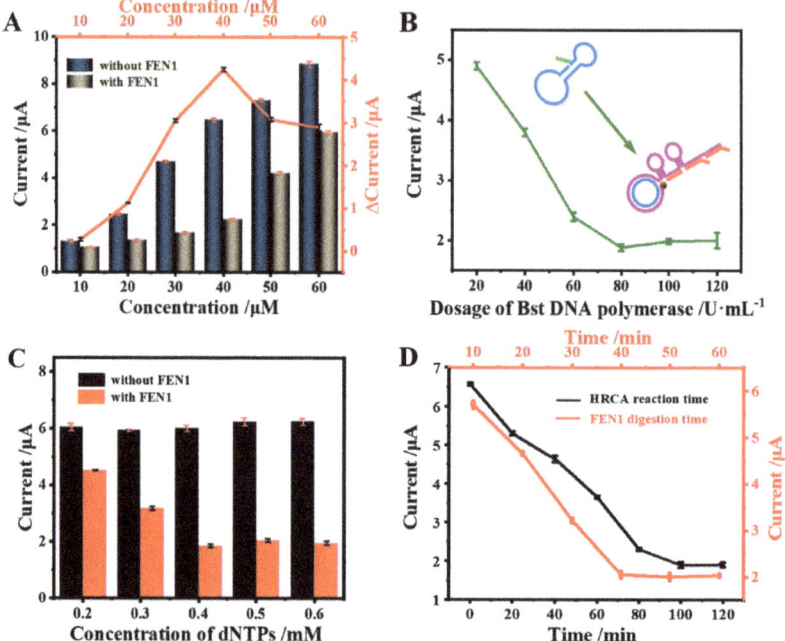

Figure 2. Condition optimization: (**A**) concentration of MB; (**B**) dosage of Bst DNA polymerase; (**C**) concentration of dNTPs; (**D**) HRCA reaction time and FEN1 digestion time. The concentrations of FEN1 and FDP are 0.5 U/L and 5 nM, respectively.

3.4. Performance of the Developed Homogeneous Electrochemical Biosensor

Under the optimal conditions, the analytical performance of this homogeneous electrochemical biosensor was evaluated by testing a series of samples containing different concentrations of FEN1. From Figure 3A, it is observed that, with the raising target concentration, the current intensity of the testing systems decreases synchronously. The inset of Figure 3A showed that there is a linear relationship between the ∆*Current* and the logarithmic concentration of FEN1 from 0.04 to 80.0 U/L. The fitted linear equation is as follows:

$$\Delta Current = 1.47 \lg C_{FEN1} + 2.85 \qquad R^2 = 0.996$$

where C represents the concentration of FEN1 (U/L), R is the meaning of the correlation linear coefficient, and ∆*Current* is expressed in units of µA. The limit of detection (LOD) of this biosensor is estimated to be 0.003 U/L (S/N = 3). As compared with the reported sensors for FEN1 detection [12,13], this designed biosensor has a wide dynamic range, higher sensitivity, and lower LOD since the HRCA reaction has greatly amplified the detecting signals. That is to say, this designed homogeneous electrochemical biosensor has a desirable linear range and high sensitivity, showing great potential in assessing FEN1 activity.

Satisfactory selectivity and anti-interference are prerequisites for the practical applicability of a sensing platform. The selectivity of this homogeneous electrochemical biosensor was verified by choosing acetylcholinesterase (AChE), Dam methyltransferase (DAM), glutathione (GSH), and bovine serum albumin (BSA) as interferences. Here, the interferences concentration was set to 100 U/L, BSA was set to 50 µg/mL, and FEN1 concentration was set to 8 U/L. None of these interferences can cleave 5′-flap from the FDP substrate to generate a dumbbell-shaped padlock probe, and thus, no HRCA can be initiated. As a result, as shown in Figure 3B, the DPV signal changes (∆*Current*) after adding these interferences are much smaller than significant signal changes upon the addition of the FEN1. These results suggest that this designed biosensor has good specificity for FEN1 detection and prominent selectivity towards its analogs.

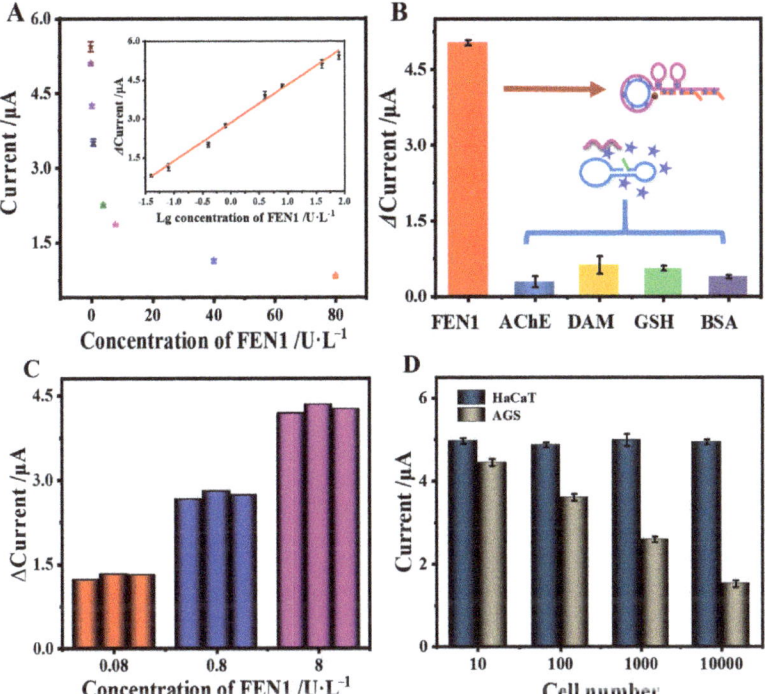

Figure 3. (**A**) Relationship between DPV peak current and FEN1 concentration. Insert: Calibration curve between $\Delta Current$ and the logarithm of the target FEN1 concentration. (**B**) Selectivity of the proposed homogeneous electrochemical biosensor in the presence of 8 U/L FEN1 and other interferences. (**C**) Reproducibility of the proposed electrochemical biosensor. (**D**) Test in the cell lysates of HaCaT and AGS cells by the proposed sensor. The error bars show the standard deviation of three replicate determinations.

The reproducibility of this proposed electrochemical biosensor was evaluated by repeated assays three times at 0.08 U/L, 0.8 U/L, and 8 U/L, respectively (see Figure 3C). The root square deviation (RSD) of each group (n = 3) at the equal concentration is 4.3%, 2.6%, and 1.9%, respectively, indicating that this proposed biosensor has good reproducibility. In addition, stability is also an important indicator to confirm the excellent performance of biosensors. Five treated ITO electrodes were used to detect the same sample (2.0 U/L FEN1) to obtain the stability curves, as shown in Figure S2. The RSD of current intensity is 2.6% (n = 5), meaning that this electrochemical biosensor has satisfactory stability. These test results verify the superior performance of this proposed electrochemical sensing platform.

3.5. Application of Biosensor to Detect FEN1 in Practical Samples

To assess the capability of the designed biosensor in practical application, this electrochemical biosensor was further applied to evaluate the FEN1 levels in cell lysates. Two cell lines, i.e., AGS (the FEN1 level positively correlated with the progression of gastric cancer) and HaCaT (the FEN1 level barely changed) were selected. As shown in Figure 3D, a significant current signal drop has been observed after adding the AGS cell lysates in the range of 10~10^4 cells, while the current scarcely decreases upon adding the HaCaT cell lysates with the equivalent cell number. This indicates that the proposed homogeneous biosensor has a satisfactory performance in the practical evaluation of FEN1 activity.

4. Conclusions

In this study, a label-free homogeneous electrochemical biosensor with ultra-high sensitivity and selectivity was proposed for FEN1 activity detection based on the HRCA technology. The well-designed FDP with 5′-flaps can be cleaved by FEN1 to form the dumbbell DNA probes, which can trigger HRCA reactions to generate dsDNA sequences. The resulting dsDNA could hinder the MB indicators to diffuse onto the IPO electrode surface, which leads to a decrease in electrochemical signals. The signal change had a direct relationship with the FEN1 concentration. Moreover, this biosensor is capable of quantifying the FEN1 activity in cell lysates with satisfactory results, indicating that this proposed electrochemical strategy has the potential to serve as a new method for FEN1 clinical assay.

Supplementary Materials: The following supporting information can be downloaded at: https://www.mdpi.com/article/10.3390/bios12070528/s1, The details of materials and reagents, cell culture and protein extraction; Figure S1: Zeta potential characterization of this sensor; Figure S2: Stability of the proposed biosensor.

Author Contributions: Writing—original draft preparation, investigation, J.Z.; data curation, X.X.; formal analysis, H.Z.; validation, Z.P.; funding acquisition, writing—review and editing, supervision, X.L.; writing—review and editing, F.L.; Conceptualization, funding acquisition, writing—review and editing, Z.L. All authors have read and agreed to the published version of the manuscript.

Funding: This research was funded by National Natural Science Foundation of China (22174018) and the Natural Science Foundation of Fujian Province (2020J01650).

Institutional Review Board Statement: Not applicable.

Informed Consent Statement: Not applicable.

Conflicts of Interest: The authors declare no conflict of interest.

References

1. Wang, K.; Xie, C.; Chen, D. Flap endonuclease 1 is a promising candidate biomarker in gastric cancer and is involved in cell proliferation and apoptosis. *Int. J. Mol. Med.* **2014**, *33*, 1268–1274. [CrossRef] [PubMed]
2. Abdel-Fatah, T.M.A.; Russell, R.; Albarakati, N.; Maloney, D.J.; Dorjsuren, D.; Rueda, O.M.; Moseley, P.; Mohan, V.; Sun, H.; Abbotts, R.; et al. Genomic and protein expression analysis reveals flap endonuclease 1 (FEN1) as a key biomarker in breast and ovarian cancer. *Mol. Oncol.* **2014**, *8*, 1326–1338. [CrossRef]
3. He, L.; Zhang, Y.; Sun, H.; Jiang, F.; Yang, H.; Wu, H.; Zhou, T.; Hu, S.; Kathera, C.S.; Wang, X.; et al. Targeting DNA Flap Endonuclease 1 to Impede Breast Cancer Progression. *EBioMedicine* **2016**, *14*, 32–43. [CrossRef] [PubMed]
4. Ma, L.; Cao, X.; Wang, H.; Lu, K.; Wang, Y.; Tu, C.; Dai, Y.; Meng, Y.; Li, Y.; Yu, P.; et al. Discovery of Myricetin as a Potent Inhibitor of Human Flap Endonuclease 1, Which Potentially Can Be Used as Sensitizing Agent against HT-29 Human Colon Cancer Cells. *J. Agric. Food Chem.* **2019**, *67*, 1656–1665. [CrossRef] [PubMed]
5. Li, S.; Jiang, Q.; Liu, Y.; Wang, W.; Yu, W.; Wang, F.; Liu, X. Precision Spherical Nucleic Acids Enable Sensitive FEN1 Imaging and Controllable Drug Delivery for Cancer-Specific Therapy. *Anal. Chem.* **2021**, *93*, 11275–11283. [CrossRef] [PubMed]
6. Wu, H.; Yan, Y.; Yuan, J.; Luo, M.; Wang, Y. miR-4324 inhibits ovarian cancer progression by targeting FEN1. *J. Ovarian Res.* **2022**, *15*, 32. [CrossRef]
7. Al-Kawaz, A.; Miligy, I.M.; Toss, M.S.; Mohammed, O.J.; Green, A.R.; Madhusudan, S.; Rakha, E.A. The prognostic significance of Flap Endonuclease 1 (FEN1) in breast ductal carcinoma in situ. *Breast Cancer Res. Tr.* **2021**, *188*, 53–63. [CrossRef]
8. Xu, L.; Shen, J.M.; Qu, J.L.; Song, N.; Che, X.F.; Hou, K.Z.; Shi, J.; Zhao, L.; Shi, S.; Liu, Y.P.; et al. FEN1 is a prognostic biomarker for ER+ breast cancer and associated with tamoxifen resistance through the ERalpha/cyclin D1/Rb axis. *Ann. Transl. Med.* **2021**, *9*, 258. [CrossRef]
9. Zhang, H.; Ba, S.; Mahajan, D.; Lee, J.Y.; Ye, R.; Shao, F.; Lu, L.; Li, T. Versatile Types of DNA-Based Nanobiosensors for Specific Detection of Cancer Biomarker FEN1 in Living Cells and Cell-Free Systems. *Nano Lett.* **2018**, *18*, 7383–7388. [CrossRef]
10. Wang, C.; Zhang, D.; Tang, Y.; Wei, W.; Liu, Y.; Liu, S. Label-Free Imaging of Flap Endonuclease 1 in Living Cells by Assembling Original and Multifunctional Nanoprobe. *ACS Appl. Bio Mater.* **2020**, *3*, 4573–4580. [CrossRef]

11. Li, B.; Zhang, P.; Zhou, B.; Xie, S.; Xia, A.; Suo, T.; Feng, S.; Zhang, X. Fluorometric detection of cancer marker FEN1 based on double-flapped dumbbell DNA nanoprobe functionalized with silver nanoclusters. *Anal. Chim. Acta* **2021**, *1148*, 238194. [CrossRef] [PubMed]
12. Tang, Y.; Wei, W.; Liu, Y.; Liu, S. Fluorescent Assay of FEN1 Activity with Nicking Enzyme-Assisted Signal Amplification Based on ZIF-8 for Imaging in Living Cells. *Anal. Chem.* **2021**, *93*, 4960–4966. [CrossRef] [PubMed]
13. Yang, H.; Wang, C.; Xu, E.; Wei, W.; Liu, Y.; Liu, S. Dual-Mode FEN1 Activity Detection Based on Nt.BstNBI-Induced Tandem Signal Amplification. *Anal. Chem.* **2021**, *93*, 6567–6572. [CrossRef] [PubMed]
14. Li, B.; Xia, A.; Xie, S.; Lin, L.; Ji, Z.; Suo, T.; Zhang, X.; Huang, H. Signal-Amplified Detection of the Tumor Biomarker FEN1 Based on Cleavage-Induced Ligation of a Dumbbell DNA Probe and Rolling Circle Amplification. *Anal. Chem.* **2021**, *93*, 3287–3294. [CrossRef]
15. Zhou, B.; Lin, L.; Li, B. Exponential amplification reaction-based fluorescent sensor for the sensitive detection of tumor biomarker flap endonuclease 1. *Sens. Actuators B Chem.* **2021**, *346*, 130457. [CrossRef]
16. Li, X.; Huang, Y.; Chen, J.; Zhuo, S.; Lin, Z.; Chen, J. A highly sensitive homogeneous electrochemiluminescence biosensor for flap endonuclease 1 based on branched hybridization chain reaction amplification and ultrafiltration separation. *Bioelectrochemistry* **2022**, *147*, 108189. [CrossRef]
17. Kimmel, D.W.; LeBlanc, G.; Meschievitz, M.E.; Cliffel, D.E. Electrochemical Sensors and Biosensors. *Anal. Chem.* **2012**, *84*, 685–707. [CrossRef]
18. Maduraiveeran, G.; Sasidharan, M.; Ganesan, V. Electrochemical sensor and biosensor platforms based on advanced nanomaterials for biological and biomedical applications. *Biosen. Bioelectron.* **2018**, *103*, 113–129. [CrossRef]
19. Xuan, F.; Fan, T.W.; Hsing, I.M. Electrochemical Interrogation of Kinetically-Controlled Dendritic DNA/PNA Assembly for Immobilization-Free and Enzyme-Free Nucleic Acids Sensing. *ACS Nano* **2015**, *9*, 5027–5033. [CrossRef]
20. Hou, T.; Xu, N.; Wang, W.; Ge, L.; Li, F. Truly Immobilization-Free Diffusivity-Mediated Photoelectrochemical Biosensing Strategy for Facile and Highly Sensitive MicroRNA Assay. *Anal. Chem.* **2018**, *90*, 9591–9597. [CrossRef]
21. Chang, J.; Wang, X.; Wang, J.; Li, H.; Li, F. Nucleic Acid-Functionalized Metal-Organic Framework-Based Homogeneous Electrochemical Biosensor for Simultaneous Detection of Multiple Tumor Biomarkers. *Anal. Chem.* **2019**, *91*, 3604–3610. [CrossRef] [PubMed]
22. Gill, R.; Patolsky, F.; Katz, E.; Willner, I. Electrochemical Control of the Photocurrent Direction in Intercalated DNA/CdS Nanoparticle Systems. *Angew. Chem. Int. Ed.* **2005**, *44*, 4554–4557. [CrossRef] [PubMed]
23. Hou, T.; Li, W.; Liu, X.; Li, F. Label-Free and Enzyme-Free Homogeneous Electrochemical Biosensing Strategy Based on Hybridization Chain Reaction: A Facile, Sensitive, and Highly Specific MicroRNA Assay. *Anal. Chem.* **2015**, *87*, 11368–11374. [CrossRef] [PubMed]
24. Thomas, D.C.; Nardone, G.A.; Randall, S.K. Amplification of Padlock Probes for DNA Diagnostics by Cascade Rolling Circle Amplification or the Polymerase Chain Reaction. *Arch. Pathol. Lab. Med.* **1999**, *123*, 1170–1176. [CrossRef] [PubMed]
25. Mohsen, M.G.; Kool, E.T. The Discovery of Rolling Circle Amplification and Rolling Circle Transcription. *Acc. Chem. Res.* **2016**, *49*, 2540–2550. [CrossRef] [PubMed]

Article

Biosensor Based on Covalent Organic Framework Immobilized Acetylcholinesterase for Ratiometric Detection of Carbaryl

Ying Luo, Na Wu, Linyu Wang, Yonghai Song, Yan Du and Guangran Ma *

National Engineering Research Center for Carbohydrate Synthesis, Key Lab of Fluorine and Silicon for Energy Materials and Chemistry of Ministry of Education, College of Chemistry and Chemical Engineering, Jiangxi Normal University, Nanchang 330022, China
* Correspondence: grma@jxnu.edu.cn or guangranma2008@163.com; Tel.: +86-0791-88120861

Abstract: A ratiometric electrochemical biosensor based on a covalent organic framework (COF$_{Thi-TFPB}$) loaded with acetylcholinesterase (AChE) was developed. First, an electroactive COF$_{Thi-TFPB}$ with a two-dimensional sheet structure, positive charge and a pair of inert redox peaks was synthesized via a dehydration condensation reaction between positively charged thionine (Thi) and 1,3,5-triformylphenylbenzene (TFPB). The immobilization of AChE on the positively charged electrode surface was beneficial for maintaining its bioactivity and achieving the best catalytic effect; therefore, the positively charged COF$_{Thi-TFPB}$ was an appropriate support material for AChE. Furthermore, the COF$_{Thi-TFPB}$ provided a stable internal reference signal for the constructed AChE inhibition-based electrochemical biosensor to eliminate various effects which were unrelated to the detection of carbaryl. The sensor had a linear range of 2.2–60 µM with a detection limit of 0.22 µM, and exhibited satisfactory reproducibility, stability and anti-interference ability for the detection of carbaryl. This work offers a possibility for the application of COF-based materials in the detection of low-level pesticide residues.

Keywords: carbaryl; acetylcholinesterase; covalent organic framework; inhibition-based electrochemical biosensor

1. Introduction

Carbaryl as a kind of carbamate pesticide has been widely applied in agricultural products due to their short-term toxicity and high insecticidal activity [1–4]. However, because of the bioaccumulation effect, a large amount of residual carbaryl in water, soil, food and the environment enter the human body through the skin, respiratory tract, digestive tract, etc., resulting in irreversible damage [5–8]. Therefore, it is crucial to achieve rapid detection and reliable quantification of carbaryl. Traditional detection methods, such as chromatography [9–12], surface-enhanced Raman spectroscopy [13–15], immunoassay [16,17], etc., have been developed very well, with high sensitivity and accuracy in the determination of pesticide residues in water and agricultural products. However, the limitations lie in the complex sample handling process, the use of highly toxic organic solvents and the expensive and complex instruments that require professional testing [18–20].

Nowadays, acetylcholinesterase (AChE) inhibition-based electrochemical biosensors have attracted great attention with regard to the detection of carbaryl and other carbamate pesticides due to their advantages of non-toxicity, simplicity, miniaturized, high specificity and high sensitivity [21–23]. For example, Loguercio et al. established the biosensor for the detection of carbaryl by immobilizing AChE on the polypyrrole nanocomposite, showing satisfactory results [24]. Zhang et al. used graphene as the support material to load AChE, and the constructed sensor realized the chiral recognition of (+)/(−)-methamidophos [25]. Considering that acetic acid, the hydrolysate of acetylthiocholine (ATCh), can induce the collapse of unstable metal-organic frameworks (MOFs), Li et al. prepared a biodegradable

ZIF-8/MB composite using the one-pot method and realized ultrasensitive detection of paraoxon by using AChE as the recognition molecule [26]. The mechanism of the AChE inhibition-based electrochemical biosensors is that the carbamate pesticides make the serine residue hydroxyl group of the AChE catalytic center be carbamylated, resulting in a decrease in its activity or even making it completely inactive. Thus, the catalytic hydrolysis of ATCh is weakened, leading to a decrease in the production of thiocholine (TCh). The response current of TCh is inversely proportional to the concentration of carbamate pesticides such as carbaryl; therefore, this causes the simple, efficient and sensitive detection of carbaryl [27–31].

The following two points are of great significance for the construction of high-performance AChE inhibition-based pesticide electrochemical sensors. Firstly, choose an appropriate support material to immobilize the enzyme and maintain its activity [32–34]. Suitable electrode support materials should allow a large number of enzymes to be loaded and provide a good microenvironment for maintaining enzyme activity [35–37]. Yang et al. reported that fixing AChE on the support material with a positive charge not only favors the maintenance of its bioactivity, but also promotes electron transport between the AChE and electrode surface [38]. Secondly, the influence of the background current and changeable environmental conditions on sensor performance should be avoided [39–41]. The traditional single-signal electrochemical sensors have low accuracy and sensitivity and poor reproducibility because their electrochemical signal is easily affected by the background current of the workstation and environmental conditions such as temperature and pH [42,43]. Fortunately, dual-signal ratiometric electrochemical sensors have emerged [44–46]. Wang et al. coated an electroactive covalent organic framework ($COF_{Thi-TFPB}$) on the surface of carbon nanotubes (CNTs) using the one-pot method, and the prepared $COF_{Thi-TFPB}$-CNT nanocomposite was used for electrochemical ratiometric detection of AA. Since the monomer thionine (Thi) was positively charged, the positively charged $COF_{Thi-TFPB}$ could self-peel into large-sized two-dimensional crystal nanosheets, and a pair of redox peaks of the $COF_{Thi-TFPB}$ was inert to the detection of AA; thus, it could be used as a reference signal [47].

Here, an electrochemical sensor based on the $COF_{Thi-TFPB}$ [48–51] loaded with AChE for the detection of carbaryl is proposed. The positively charged $COF_{Thi-TFPB}$ can be easily stripped into two-dimensional nanosheets, and modified on the surface of a bare glassy carbon electrode (GCE) without adhesives or conductive agents. The positively charged COFThi−TFPB is conducive to effectively maintaining the bioactivity of AChE and achieving the best catalytic effect. Its inherent redox peak at $0/-0.22$ V is inert to the detection of carbaryl, which could be used as a reference signal to further improve the sensitivity and accuracy of the detection. The prepared sensor shows good reproducibility, stability and anti-interference ability. This work proposes an efficient strategy to immobilize enzymes using an electroactive COF as a support material.

2. Experimental Procedure

2.1. Materials and Reagents

1,3,5-triformylphenylbenzene (TFPB) and thionine (Thi) were purchased from Jilin Yanshen Technology Co., Ltd., (Beijing, China). N,N-dimethylformamide (DMF), N,N-dimethylacetamide (DMA), mesitylene, tetrahydrofuran (THF), acetic acid (AcOH), acetylcholinesterase (AChE), acetylthiocholine (ATCh) and other chemicals were purchased from Inokay Co., Ltd. (Beijing, China). Carbaryl was purchased from Mokai Nike Technology Co., Ltd. (Jiangxi, China).

2.2. Instruments

Scanning/transmission electron microscopy images (SEM/TEM) were obtained via the HITACHI S-3400N SEM and JM-2010 (HR) TEM (Chiyoda City, Japan), respectively. Atomic force microscopy (AFM) images were obtained via the instrument model BRUKER Nanoscope V (MultiMode 8) (Billerica, MA, USA) multifunctional scanning probe microscope. Fourier transform infrared spectroscopy (FTIR) was recorded on model Perkin-Elmer Spectrum 100 spectrometer (Waltham, MA, USA). N_2 adsorption/desorption isotherm measurements were operated using a BELSORP-mini II instrument (Microtrac, Haan/Duesseldorf, Germany) under the liquid nitrogen temperature of 77 K. Powder X-ray diffraction (XRD) analysis was performed on the D/Max 2500 V/PC X-ray powder diffractometer (Rigaku, Tokyo, Japan) with a scanning step of 1°/min. All electrochemical studies were performed on an electrochemical workstation (CHI 760D, Shanghai, China).

2.3. Preparation of $COF_{Thi-TFPB}$

Firstly, 0.2 mM TFPB and 0.3 mM Thi were added to a mixture with 2 mL of 1,4-dioxane, and 1 mL of mesitylene and DMF, and ultrasound was performed for 15 min. Next, it was transferred to a 25 mL reaction kettle with 0.2 mL (concentration: 6 M) acetic acid (used as an initiator), and placed in an oven at 120 °C for three days. Finally, the dark-blue $COF_{Thi-TFPB}$ was obtained by centrifugation and freeze-drying [52].

2.4. Preparation of $AChE/COF_{Thi-TFPB}/GCE$

Firstly, the surface of the glassy carbon electrode (GCE) was treated with Al_2O_3, ethanol and ultrapure water until smooth and clean. Then, 5 μL of the 2 mg/mL $COF_{Thi-TFPB}$ was dropped on the electrode surface. After drying, $AChE/COF_{Thi-TFPB}/GCE$ was prepared by dropping 0.4 mM AChE on the modified electrode. The detection mechanism of carbaryl is shown in Scheme 1.

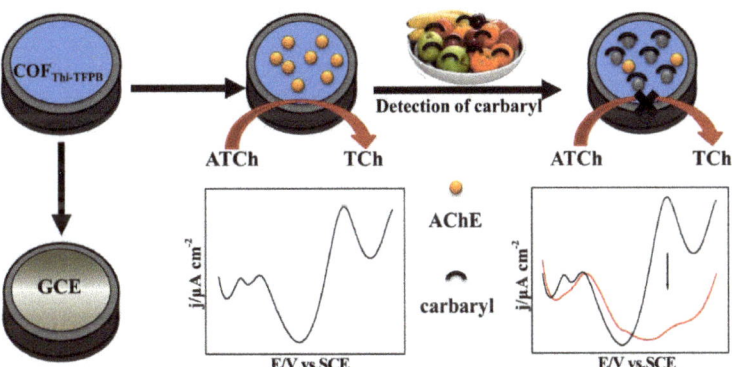

Scheme 1. Schematic illustration for the detection mechanism of carbaryl.

3. Results and Discussion

3.1. Characterization of $COF_{Thi-TFPB}$

SEM (Figure 1a), TEM (Figure 1b) and AFM (Figure 1c) showed that the $COF_{Thi-TFPB}$ owned a film-like lamellar structure, and the thickness was about 1.42 nm, which was very favorable for the immobilization of AChE [53,54]. Next, an FTIR spectrum and XRD pattern were used to demonstrate the successful synthesis of the $COF_{Thi-TFPB}$. The spectrum of the $COF_{Thi-TFPB}$ showed that a new peak of -C=N- appeared at 1658 cm^{-1}, whereas the disappearance of N—H in –NH_2 at 3292 cm^{-1} and C=O in –CHO at 1689 cm^{-1}. Meanwhile, the stretching vibration peaks corresponding to –CH in –CHO at 2714 cm^{-1} and 2834 cm^{-1} disappeared. These results demonstrated that the $COF_{Thi-TFPB}$ was synthesized successfully (Figure 1d). The XRD pattern further confirmed the successful synthesis of the crystalline $COF_{Thi-TFPB}$ (Figure 1e). Diffraction peaks at 6.67°, 8.48°, 11.34°, 25.8° and 45.6° corresponded to (100), (110), (210), (152) and (111) crystal planes. The N_2 adsorption/desorption isotherm of the $COF_{Thi-TFPB}$ showed that the specific surface area was 67.4 $m^2\ g^{-1}$ (Figure 1f).

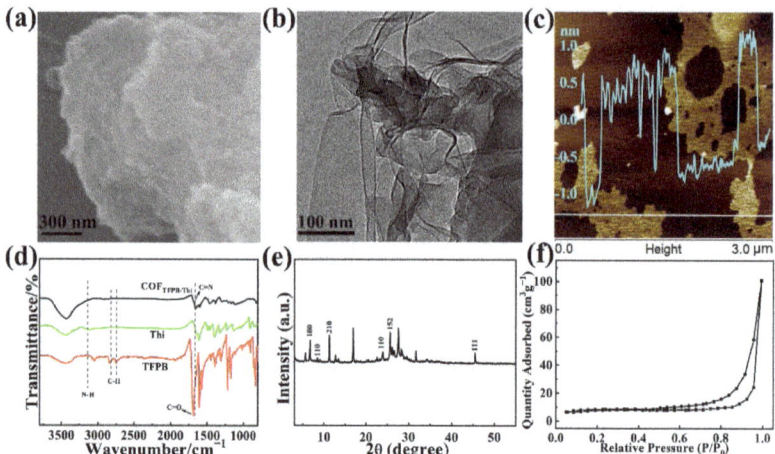

Figure 1. SEM, TEM and AFM images of COF$_{Thi-TFPB}$ (**a–c**). FTIR spectrum (**d**), XRD pattern (**e**) and N$_2$ adsorption/desorption isotherm (**f**) of COF$_{Thi-TFPB}$.

3.2. Electrochemical Behaviors of COF$_{Thi-TFPB}$/GCE and AChE/COF$_{Thi-TFPB}$/GCE

To successfully construct an electrochemical sensor for ratiometric detection of carbaryl, the introduced internal reference signal should have good stability and a suitable oxidation potential. Therefore, the electrochemical performance of the COF$_{Thi-TFPB}$ was evaluated by cyclic voltammetry (CV). As shown in Figure 2a, compared with the TFPB/GCE (b) without a peak, both the COF$_{Thi-TFPB}$/GCE (c) and Thi/GCE (a) had two pairs of redox peaks at the same potential, indicating that the peaks on the COF$_{Thi-TFPB}$/GCE came from the electroactive Thi. Next, the CV of the COF$_{Thi-TFPB}$/GCE was investigated at different scan rates. It could be seen that the positions of their redox peaks were basically unchanged with the increase in scanning rate (Figure 2b). The redox peak at −0.2/−0.07 V was caused by the electrochemical reaction of Thi in the COF$_{Thi-TFPB}$, and the redox peak at 0/−0.22 V was due to the conjugated structure of Thi in the COF$_{Thi-TFPB}$ [55]. With the increase in scanning rate, the peak current density of the two redox peaks of the COF$_{Thi-TFPB}$ also increased and showed a good linear relationship, indicating that the reaction process was a typical surface control process (Figure 2c). Based on the linear regression equation between the anodic/cathodic peak potential and natural logarithm of the scan rate (Figure 2d), it could be calculated that the electron-transfer number (n) was 1, and the electron-transfer coefficient (αs) was 0.369 when the redox potential was 0/−0.22 V [56].

Then, CV and electrochemical impedance spectroscopy (EIS) tests were performed using 5.0 mM [Fe(CN)$_6$]$^{3-/4-}$ in a 0.1 M KCl solution as a probe to investigate the electrochemical behaviors of the COF$_{Thi-TFPB}$/GCE and AChE/COF$_{Thi-TFPB}$/GCE. As shown in Figure 3a, compared with the bare GCE (curve a), the peak current of [Fe(CN)$_6$]$^{3-/4-}$ on the COF$_{Thi-TFPB}$/GCE (curve b) was slightly increased, and the peak-to-peak potential difference was slightly decreased. This result might be attributed to the electrostatic attraction between the negative [Fe(CN)$_6$]$^{3-/4-}$ and the positive COF$_{Thi-TFPB}$. In the meantime, the redox peak of [Fe(CN)$_6$]$^{3-/4-}$ on the AChE/COF$_{Thi-TFPB}$/GCE (curve c) owned the smallest peak current and the largest peak-to-peak potential difference. It was mainly due to the poor conductivity of AChE, which would hinder the electron transport between [Fe(CN)$_6$]$^{3-/4-}$ and the surface of the electrode [57,58]. The charge transfer resistance (R_{ct}) values of the bare GCE, COF$_{Thi-TFPB}$/GCE and AChE/COF$_{Thi-TFPB}$/GCE were 31.3 Ω, 8.5 Ω and 288.1 Ω, respectively (Figure 3b). In conclusion, compared with the COF$_{Thi-TFPB}$/GCE, the peak current of [Fe(CN)$_6$]$^{3-/4-}$ on the AChE/COF$_{Thi-TFPB}$/GCE decreased and the R_{ct} value increased, which directly proved the successful fixation of AChE on the COF$_{Thi-TFPB}$/GCE.

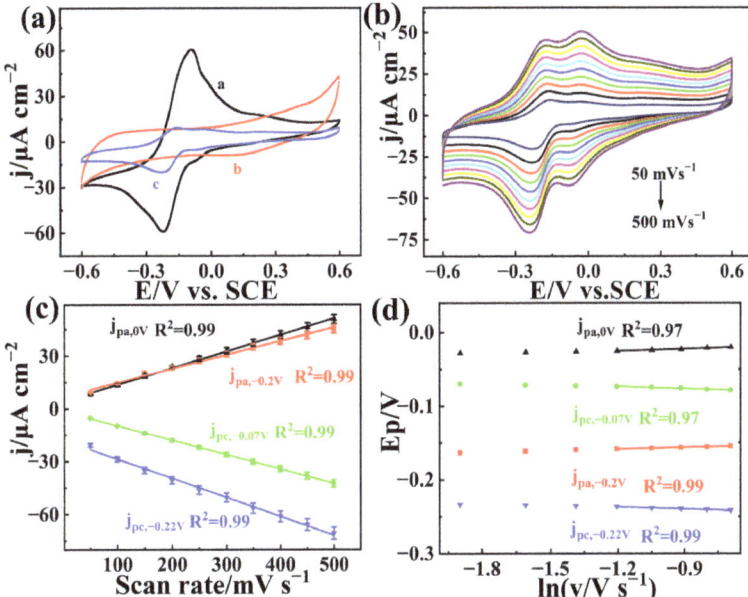

Figure 2. (**a**) CV of TFPB/GCE (curve b), Thi/GCE (curve a) and COF$_{Thi\text{-}TFPB}$/GCE (curve c) in 0.1 M N$_2$-statured PBS (pH = 7.0). (**b**) CVs of COF$_{Thi\text{-}TFPB}$/GCE at different scan rates (50 mV s^{-1} to 500 mV s^{-1}) in 0.1 M N$_2$-statured PBS (pH = 7.0). (**c**) The corresponding fitting curves between current density and scan rates. (**d**) The corresponding fitting curves between Ep and ln υ.

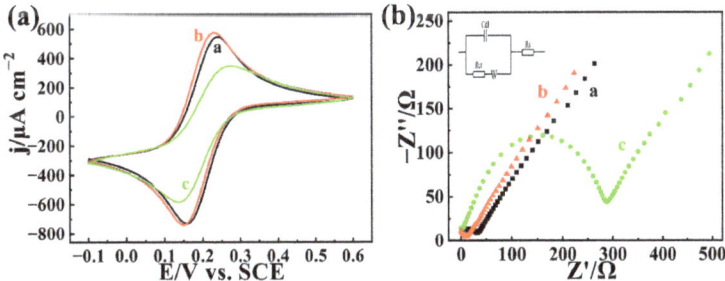

Figure 3. (**a**) CVs and (**b**) EIS of bare/GCE (curve a) and COF$_{Thi\text{-}TFPB}$/GCE (curve b), AChE/COF$_{Thi\text{-}TFPB}$/GCE (curve c) in 0.1 M KCl solution with 5.0 mM [Fe(CN)$_6$]$^{3-/4-}$. Inset in b is its equivalent circuit.

3.3. Optimization of the Experimental Conditions

It was known that when at the optimum pH value, the binding ability of the enzyme molecule to the substrate was the strongest and the enzyme reaction rate was the highest; however, if the pH was too large or too small, the enzyme might be inactivated [59,60]. Therefore, the pH value of the solution was optimized. As shown in Figure 4a, when pH = 7.0, the current density of the AChE/COF$_{Thi\text{-}TFPB}$/GCE was the largest in the 0.1 M PBS with 0.6 mM ATCh and 5 μM carbaryl. Then, the amount of the COF$_{Thi\text{-}TFPB}$ modified on the electrode surface was optimized (Figure 4b), which showed that the optimal volume was 5 μL (concentration: 2 mg/mL). Next, considering that the amount of loaded AChE and the concentration of substrate molecule ATCh had important influence on the detection results, they were also optimized. It could be observed that it was best to set the concentration of AChE and ATCh at 0.4 mM and 0.6 mM in the subsequent experiments, respectively (Figure 5a,b). Figure 5c showed the relationship between the catalytic activity inhibition of AChE and the incubation time. The inset is the formula for calculating the percentage of inhibition ($I\%$), where $j_{P,control}$ was the original current density recorded by the AChE/COF$_{Thi-TFPB}$/GCE in 0.1 M PBS (pH = 7.0) with 0.6 mM ATCh, $j_{P,exp}$ was the residual current density recorded after immersing in 0.1 M PBS (pH = 7.0) with 0.6 mM ATCh and 5 μM carbaryl for 0, 4, 8, 12, 16, 20 and 30 min. All in all, when the concentration of AChE was 0.4 mM, ATCh was 0.6 mM and the incubation time was 20 min, the performance of the AChE/COF$_{Thi-TFPB}$/GCE sensor was the best.

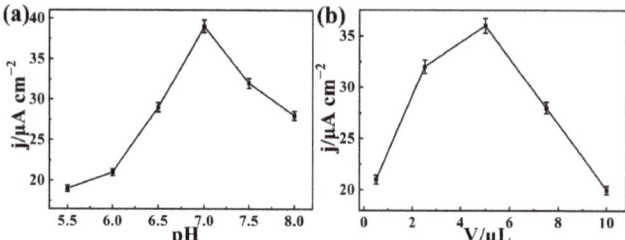

Figure 4. The plots of peak current density versus different pH (**a**) and different volumes of COF$_{Thi-TFPB}$ (**b**).

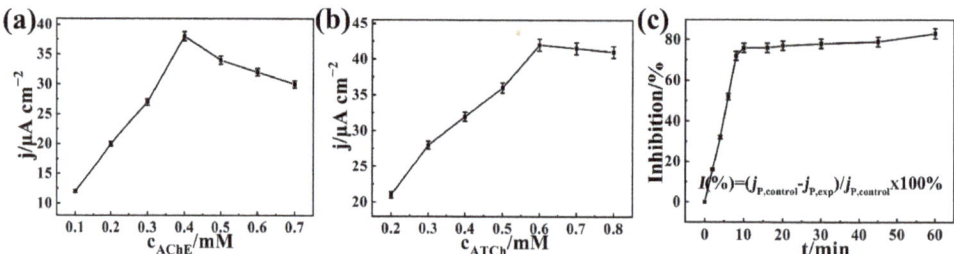

Figure 5. The plots of peak current density versus different concentrations of AChE (**a**), different concentrations of ATCh (**b**) and different incubation times (**c**).

3.4. Electrochemical Detection of Carbaryl Based on AChE/COF$_{Thi-TFPB}$/GCE

Firstly, the affinity of AChE fixed on the COF$_{Thi-TFPB}$ to ATCh was investigated. Figure 6a shows the relation curve between response current density and time after continuously adding ATCh. It could be seen that there was a good linear relationship between the oxidation peak current density and the concentration of ATCh between 0.01 mM and 0.27 mM. However, the slow-response current density at higher concentrations of ATCh indicated a Michaelis–Menten process. According to the slope and intercept of the linear regression equation in Figure 6b, the Michaelis–Menten constant (Km) was calculated to be 0.24 mM. This value was lower than 0.622 mM as measured by the AChE/COF@MWCNTs/GCE [61], suggesting good affinity between the enzyme and the substrate. Then, a ratiometric electrochemical sensor based on the AChE/COF$_{Thi-TFPB}$ was used to detect carbaryl in 0.1 M PBS (pH = 7.0) containing 0.6 mM ATCh. As shown in Figure 7a, the peak current density of the COF$_{Thi-TFPB}$ at -0.05 V was basically unchanged with the addition of carbaryl, whereas the peak current density of TCh at 0.6 V gradually decreased. This was because the toxic effect of carbaryl on AChE led to a decrease in the amount of TCh, and then, the electrochemical signal was weakened. The inset in Figure 7a shows the linear relationship between j_{TCh}/j_{COF} and the concentration of carbaryl, where the linear range of the carbaryl sensor was 2.2–60 μM and the detection limit was 0.22 μM. The performance of the AChE/COF$_{Thi-TFPB}$/GCE sensor was compared with other sensors (Table 1). It could be seen that the detection limit of this sensor was lower than that based on Au/PAMAM/GLUT/AChE (3.2 μM) and MWCNT/PANI/AChE (1.4 μM).

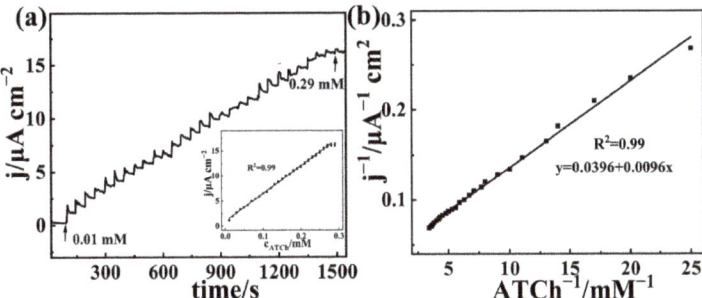

Figure 6. (a) Amperometric response current density for AChE/COF$_{Thi\text{-}TFPB}$/GCE obtained by continuously adding ATCh in 0.1 M PBS (pH = 7.0) with constant stirring at a voltage set to 0.7 V. Inset: the plots of response current density versus different concentrations of ATCh. (b) The Lineweaver–Burk plot.

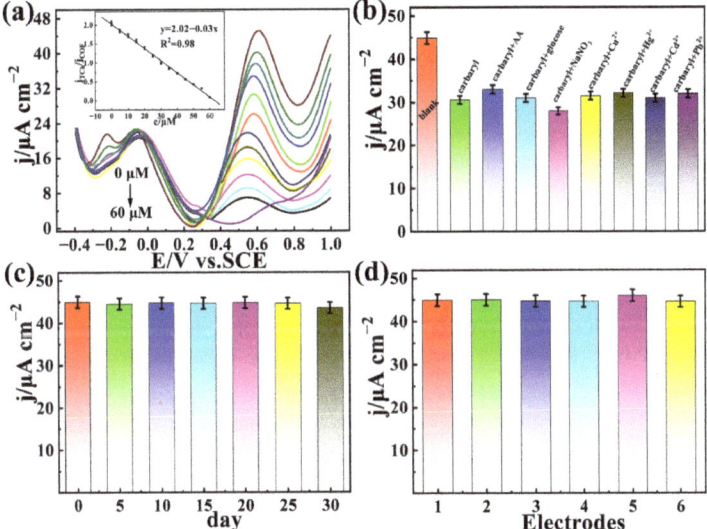

Figure 7. (a) The DPV response of AChE/COF$_{Thi\text{-}TFPB}$/GCE in 0.1 M PBS (pH = 7.0) containing 0.6 mM ATCh with different concentrations of carbaryl. (Inset: fitting curve between current density and the concentration of carbaryl). (b) The selectivity of AChE/COF$_{Thi\text{-}TFPB}$/GCE in 0.1 M PBS (pH = 7.0) with 0.6 mM ATCh, 10 μM carbaryl and 50 μM interferences. (c,d) The stability and reproducibility of AChE/COF$_{Thi\text{-}TFPB}$/GCE in 0.1 M PBS (pH = 7.0) with 0.6 mM ATCh and 5 μM carbaryl.

The selectivity of the AChE/COF$_{Thi\text{-}TFPB}$/GCE sensor was investigated in 0.1 M PBS (pH = 7.0) with 0.6 mM ATCh, 10 μM carbaryl and 50 μM interferences. It could be seen that these interferences had little effect on the peak current density (Figure 7b). Then, one AChE/COF$_{Thi\text{-}TFPB}$/GCE was used to measure the corresponding peak current density of 5 μM carbaryl in 0.1 M PBS (pH = 7.0) with 0.6 mM ATCh for 30 days. The relative standard deviation (RSD) was only 1.15%, indicating that the AChE/COF$_{Thi\text{-}TFPB}$/GCE sensor had good stability (Figure 7c). The RSD of 5 μM carbaryl detected by six independent AChE/COF$_{Thi\text{-}TFPB}$/GCEs was 1.45% in 0.1 M PBS (pH = 7.0) with 0.6 mM ATCh (Figure 7d). The good stability and reproducibility might be attributed to the fact that the positively charged COF$_{Thi\text{-}TFPB}$ could immobilize the AChE enzyme and maintain its activity, and its oxidation peak played a self-correcting role in the detection of carbaryl.

Table 1. Performance comparison of several carbaryl electrochemical sensors.

Electrode	LOD (µM)	Linear Range (µM)	Reference
GC/rGO/AChE	0.0019	0.2–10	[62]
Au/PAMAM/GLUT/AChE	3.2	1–9	[63]
Nafion/AChE/CHIT/IAM	0.004	0.005–5.0	[64]
AChE–MWCNTs/GONRs/GCE	0.0017	0.005–5.0	[65]
AChE/PDDA-MWCNTs-GR/GCE	0.001	0.255–14.9	[66]
MWCNT/PANI/AChE	1.4	9.9–49.6	[67]
GO-IL/GCE	0.02	0.1–12.0	[68]
ZXCPE	0.3	1–100	[69]
CB-NP electrode	12	25–125	[70]
IL/CC	1.4	10–75	[71]
AChE/COF$_{Thi-TFPB}$/GCE	0.22	2.2–60	This work

3.5. Detection of Carbaryl in Vegetable Samples

In addition, the AChE/COF$_{Thi-TFPB}$/GCE sensor and high-performance liquid chromatography (HPLC) were used to detect carbaryl in real samples to demonstrate the practical application capability of the sensor. Firstly, a 100 g lettuce sample was chopped and put into a juicer containing 100 mL of 0.1 M PBS (pH = 7.0). Then, the obtained mixture was filtered, and the filtrate was used as the actual sample. Next, different concentrations of carbaryl were added to the actual sample, and the carbaryls in the actual samples were determined by the AChE/COF$_{Thi-TFPB}$/GCE sensor and HPLC. The obtained results are shown in Table 2. It could be seen that the carbaryl content in the actual samples obtained by the AChE/COF$_{Thi-TFPB}$/GCE sensor was close to the results of the HPLC test, which proved that the AChE/COF$_{Thi-TFPB}$/GCE sensor has the potential to detect carbaryl in real examples.

Table 2. The detection of carbaryl in lettuce juice by AChE/COF$_{Thi-TFPB}$/GCE and HPLC.

Sample	Added (µM)	Found (µM)	Average Value (µM)	Recovery (%)	RSD (%, $n = 3$)	HPLC (µM)	RSD (%, $n = 3$)
1	0	-	-	-	-	-	-
2	5	4.85, 5.03, 4.96	4.95	99	1.8	4.98	1.8
3	10	10.2, 10.8, 9.8	10.27	102.7	4.9	10.33	4.9
4	20	20.7, 20.1, 20.5	20.4	102	1.5	20.45	1.5

4. Conclusions

The development of efficient, good, stable and reproducible AChE inhibition-based electrochemical biosensors might rely on the immobilization of the enzyme on suitable support materials and the introduction of internal reference signals to eliminate irrelevant effects in detection. In this work, a ratiometric electrochemical sensor was constructed by using the positively charged COF$_{Thi-TFPB}$ with an inert redox peak as the support material to load the AChE enzyme. The COF$_{Thi-TFPB}$ could immobilize the AChE enzyme and maintain its activity. On the other hand, its inherent redox peak at 0/−0.22 V was inert to the detection of carbaryl, which could be used as a reference signal to further improve the sensitivity and accuracy of the detection. The linear range of this sensor was 2.2–60 µM, the detection limit was 0.22 µM, and it had good selectivity, reproducibility and stability. This suggests that this material has the potential to be applied to detect low-level pesticide residues.

Author Contributions: Y.L. and G.M. conceived and planned the study. Y.L., N.W. and L.W. carried out the experiments. Y.L., Y.S. and Y.D. writing—review and editing the manuscript. G.M. was responsible for supervision, project administration and funding acquisition. All authors provided critical feedback and helped shape the research, analysis. All authors have read and agreed to the published version of the manuscript.

Funding: This work was financially supported by the National Natural Science Foundation of China (21964010).

Institutional Review Board Statement: Not applicable.

Informed Consent Statement: Not applicable.

Data Availability Statement: The data is available under the request to the correspondence.

Conflicts of Interest: The authors declare no conflict of interest.

References

1. Nunes, E.; Silva, M.; Rascon, J.; Leiva-Tafur, D.; Lapa, R.; Cesarino, I. Acetylcholinesterase biosensor based on functionalized renewable carbon platform for detection of carbaryl in food. *Biosensors* **2022**, *12*, 486. [CrossRef] [PubMed]
2. Raymundo-Pereira, P.A.; Gomes, N.O.; Shimizu, F.M.; Machado, A.S.; Oliveira, O.N., Jr. Selective and sensitive multiplexed detection of pesticides in food samples using wearable, flexible glove-embedded non-Enzymatic Sensors. *Chem. Eng. J.* **2021**, *408*, 127279. [CrossRef]
3. Paschoalin, R.T.; Gomes, N.O.; Almeida, G.F.; Bilatto, S.; Farinas, C.S.; Machado, S.A.S.; Mattoso, L.H.C.; Oliveira, O.N., Jr.; Raymundo-Pereira, P.A. Wearable sensors made with solution-blow spinning poly (lactic acid) for non-enzymatic pesticide detection in agriculture and food safety. *Biosens. Bioelectron.* **2022**, *199*, 11385. [CrossRef] [PubMed]
4. Umapathi, R.; Sonwal, S.; Lee, M.; Rani, G.; Lee, E.; Jeon, T.; Kang, S.; Oh, M.; Huh, Y. Colorimetric based on-site sensing strategies for the rapid detection of pesticides in agricultural foods: New horizons, perspectives, and challenges. *Coord. Chem. Rev.* **2021**, *446*, 214061. [CrossRef]
5. Jiang, W.; Liu, Y.; Ke, Z.; Zhang, L.; Zhang, M.; Zhou, Y.; Wang, H.; Wu, C.; Qiu, J.; Hong, Q. Substrate preference of carbamate hydrolase CehA reveals its environmental behavior. *J. Hazard. Mater.* **2021**, *403*, 123677. [CrossRef]
6. Su, D.; Zhao, X.; Yan, X.; Han, X.; Zhu, Z.; Wang, C.; Jia, X.; Liu, F.; Sun, P.; Liu, X.; et al. Background-free sensing platform for on-site detection of carbamate pesticide through upconversion nanoparticles-based hydrogel suit. *Biosens. Bioelectron.* **2021**, *194*, 113598. [CrossRef]
7. Madrigal, J.; Jones, R.; Gunier, R.; Whitehead, T.; Reynolds, P.; Metayer, C.; Ward, M. Residential exposure to carbamate, organophosphate, and pyrethroid insecticides in house dust and risk of childhood acute lymphoblastic leukemia. *Environ. Res.* **2021**, *201*, 111501. [CrossRef]
8. Chen, Z.; Wu, H.; Xiao, Z.; Fu, H.; Shen, Y.; Luo, L.; Wang, H.; Lei, H.; Hongsibsong, S.; Xu, Z. Rational hapten design to produce high-quality antibodies against carbamate pesticides and development of immunochromatographic assays for simultaneous pesticide screening. *J. Hazard. Mater.* **2021**, *412*, 125241. [CrossRef]
9. Wang, S.; Shi, X.; Liu, F.; Laborda, P. Chromatographic methods for detection and quantification of carbendazim in food. *J. Agric. Food Chem.* **2020**, *68*, 11880–11894. [CrossRef]
10. Ruengprapavut, S.; Sophonnithiprasert, T.; Pongpoungphet, N. The effectiveness of chemical solutions on the removal of carbaryl residues from cucumber and chili presoaked in carbaryl using the HPLC technique. *Food Chem.* **2020**, *309*, 125659. [CrossRef]
11. Chullasat, K.; Huang, Z.; Bunkoed, O.; Kanatharana, P.; Lee, H. Bubble-in-drop microextraction of carbamate pesticides followed by gas chromatography-mass spectrometric analysis. *Microchem. J.* **2020**, *155*, 104666. [CrossRef]
12. Mao, X.; Xiao, W.; Wan, Y.; Li, Z.; Luo, D.; Yang, H. Dispersive solid-phase extraction using microporous metal-organic framework UiO-66: Improving the matrix compounds removal for assaying pesticide residues in organic and conventional vegetables. *Food Chem.* **2021**, *345*, 128807. [CrossRef] [PubMed]
13. Zhang, D.; Liang, P.; Chen, W.; Tang, Z.; Li, C.; Xiao, K.; Jin, S.; Ni, D.; Yu, Z. Rapid field trace detection of pesticide residue in food based on surface-enhanced Raman spectroscopy. *Microchim. Acta* **2021**, *188*, 370. [CrossRef] [PubMed]
14. Wang, S.; Shi, X.; Zhu, G.; Zhang, Y.; Jin, D.; Zhou, Y.; Liu, F.; Laborda, P. Application of surface-enhanced Raman spectroscopy using silver and gold nanoparticles for the detection of pesticides in fruit and fruit juice. *Trends Food Sci. Technol.* **2021**, *116*, 583–602. [CrossRef]
15. Pu, H.; Huang, Z.; Xu, F.; Sun, D.-W. Two-dimensional self-assembled Au-Ag core-shell nanorods nanoarray for sensitive detection of thiram in apple using surface-enhanced Raman spectroscopy. *Food Chem.* **2020**, *343*, 128548. [CrossRef]
16. Zha, Y.; Li, Y.; Hu, P.; Lu, S.; Ren, H.; Liu, Z.; Yang, H.; Zhou, Y. Duplex-specific nuclease-triggered fluorescence immunoassay based on dual-functionalized AuNP for acetochlor, metolachlor, and propisochlor. *Anal. Chem.* **2021**, *93*, 13886–13892. [CrossRef]
17. Zhao, Y.; Ruan, X.; Song, Y.; Smith, J.; Vasylieva, N.; Hammock, B.; Lin, Y.; Du, D. Smartphone-based dual-channel immunochromatographic test strip with polymer quantum dot labels for simultaneous detection of cypermethrin and 3-phenoxybenzoic acid. *Anal. Chem.* **2021**, *93*, 13658–13666. [CrossRef]
18. Li, Z.; Zhou, J.; Dong, T.; Xu, Y.; Shang, Y. Application of electrochemical methods for the detection of abiotic stress biomarkers in plants. *Biosens. Bioelectron.* **2021**, *182*, 113105. [CrossRef]
19. Sinha, A.; Ma, K.; Zhao, H. 2D $Ti_3C_2T_x$ flakes prepared by in-situ HF etchant for simultaneous screening of carbamate pesticides. *J. Colloid Interface Sci.* **2021**, *590*, 365–374. [CrossRef]
20. Wu, J.; Yang, Q.; Li, Q.; Li, H.; Li, F. Two-Dimensional MnO_2 nanozyme-mediated homogeneous electrochemical detection of organophosphate pesticides without the interference of H_2O_2 and color. *Anal. Chem.* **2021**, *93*, 4084–4091. [CrossRef]

21. Qi, J.; Tan, D.; Wang, X.; Ma, H.; Wan, Y.; Hu, A.; Li, L.; Xiao, B.; Lu, B. A novel acetylcholinesterase biosensor with dual-recognized strategy based on molecularly imprinted polymer. *Sens. Actuators B* **2021**, *337*, 129760. [CrossRef]
22. Fernandez-Ramos, M.; Ogunneye, A.; Babarinde, N.; Erenas, M.; Capitan-Vallvey, L. Bioactive microfluidic paper device for pesticide determination in waters. *Talanta* **2020**, *218*, 121108. [CrossRef] [PubMed]
23. Zhu, Y.; Wang, M.; Zhang, X.; Cao, J.; She, Y.; Cao, Z.; Wang, J.; Abd El-Aty, A. Acetylcholinesterase immobilized on magnetic mesoporous silica nanoparticles coupled with fluorescence analysis for rapid detection of carbamate pesticides. *ACS Appl. Nano Mater.* **2022**, *5*, 1327–1338. [CrossRef]
24. Loguercio, L.; Thesing, A.; Demingos, P.; de Albuquerque, C.; Rodrigues, R.; Brolo, A.; Santos, J. Efficient acetylcholinesterase immobilization for improved electrochemical performance in polypyrrole nanocomposite-based biosensors for carbaryl pesticide. *Sens. Actuators B* **2021**, *339*, 129875. [CrossRef]
25. Zhang, Y.; Liu, X.; Qiu, S.; Zhang, Q.; Tang, W.; Liu, H.; Guo, Y.; Ma, Y.; Guo, X.; Liu, Y. A flexible acetylcholinesterase-modified graphene for chiral pesticide sensor. *J. Am. Chem. Soc.* **2019**, *141*, 14643–14649. [CrossRef]
26. Li, X.; Gao, X.; Gai, P.; Liu, X.; Li, F. Degradable metal-organic framework/methylene blue composites-based homogeneous electrochemical strategy for pesticide assay. *Sens. Actuators B* **2020**, *323*, 128701. [CrossRef]
27. Chen, J.; Chen, X.; Wang, P.; Liu, S.; Chi, Z. Aggregation-induced emission luminogen@manganese dioxide core-shell nanomaterial-based paper analytical device for equipment-free and visual detection of organophosphorus pesticide. *J. Hazard. Mater.* **2021**, *413*, 125306. [CrossRef]
28. Montali, L.; Calabretta, M.; Lopreside, A.; D'Elia, M.; Guardigli, M.; Michelini, E. Multienzyme chemiluminescent foldable biosensor for on-site detection of acetylcholinesterase inhibitors. *Biosens. Bioelectron.* **2020**, *162*, 112232. [CrossRef]
29. Yu, L.; Chang, J.; Zhuang, X.; Li, H.; Hou, T.; Li, F. Two-dimensional cobalt-doped Ti_3C_2 MXene nanozyme-mediated homogeneous electrochemical strategy for pesticides assay based on in situ generation of electroactive substances. *Anal. Chem.* **2022**, *94*, 3669–3676. [CrossRef]
30. Zhu, H.; Liu, P.; Xu, L.; Li, X.; Hu, P.; Liu, B.; Pan, J.; Yang, F.; Niu, X. Nanozyme-participated biosensing of pesticides and cholinesterases: A critical review. *Biosensors* **2021**, *11*, 382. [CrossRef]
31. Rafat, N.; Satoh, P.; Worden, R. Electrochemical biosensor for markers of neurological esterase inhibition. *Biosensors* **2021**, *11*, 459. [CrossRef] [PubMed]
32. Pinyou, P.; Blay, V.; Muresan, L.; Noguer, T. Enzyme-modified electrodes for biosensors and biofuel cells. *Mater. Horiz.* **2019**, *6*, 1336–1358. [CrossRef]
33. He, Y.; Hu, F.; Zhao, J.; Yang, G.; Zhang, Y.; Chen, S.; Yuan, R. Bifunctional moderator-powered ratiometric electrochemiluminescence enzymatic biosensors for detecting organophosphorus pesticides based on dual-signal combined nanoprobes. *Anal. Chem.* **2021**, *93*, 8783–8790. [CrossRef] [PubMed]
34. Itsoponpan, T.; Thanachayanont, C.; Hasin, P. Sponge-like $CuInS_2$ microspheres on reduced graphene oxide as an electrocatalyst to construct an immobilized acetylcholinesterase electrochemical biosensor for chlorpyrifos detection in vegetables. *Sens. Actuators B* **2021**, *337*, 129775. [CrossRef]
35. Kadambar, V.; Bellare, M.; Bollella, P.; Katz, E.; Melman, A. Electrochemical control of the catalytic activity of immobilized enzymes. *Chem. Commun.* **2020**, *56*, 13800–13803. [CrossRef]
36. Welden, M.; Poghossian, A.; Vahidpour, F.; Wendlandt, T.; Keusgen, M.; Wege, C.; Schoning, M. Towards multi-analyte detection with field-effect capacitors modified with tobacco mosaic virus bioparticles as enzyme nanocarriers. *Biosensors* **2022**, *12*, 43. [CrossRef]
37. Arnold, J.; Chapman, J.; Arnold, M.; Dinu, C. Hyaluronic acid allows enzyme immobilization for applications in biomedicine. *Biosensor* **2022**, *12*, 28. [CrossRef] [PubMed]
38. Yang, S.; Liu, J.; Zheng, H.; Zhong, J.; Zhou, J. Simulated revelation of the adsorption behaviours of acetylcholinesterase on charged self-assembled monolayers. *Nanoscale* **2020**, *12*, 3701–3714. [CrossRef]
39. Zhang, C.; Liu, Z.; Zhang, L.; Zhu, A.; Liao, F.; Wan, J.; Zhou, J.; Tian, Y. A robust Au−C≡C functionalized surface: Toward real-time mapping and accurate quantification of Fe^{2+} in the brains of live AD mouse models. *Angew. Chem. Int. Ed.* **2020**, *59*, 20499–20507. [CrossRef]
40. Zhong, W.; Gao, F.; Zou, J.; Liu, S.; Li, M.; Gao, Y.; Yu, Y.; Wang, X.; Lu, L. MXene@Ag-based ratiometric electrochemical sensing strategy for effective detection of carbendazim in vegetable samples. *Food Chem.* **2021**, *360*, 130006. [CrossRef]
41. Yang, G.; He, Y.; Zhao, J.; Chen, S.; Yuan, R. Ratiometric electrochemiluminescence biosensor based on Ir nanorods and CdS quantum dots for the detection of organophosphorus pesticides. *Sens. Actuators B* **2021**, *341*, 130008. [CrossRef]
42. Zhang, K.; Lv, S.; Tang, D. Novel 3D printed device for dual-signaling ratiometric photoelectrochemical readout of biomarker using λ-exonuclease-assisted recycling amplification. *Anal. Chem.* **2019**, *91*, 10049–10055. [CrossRef] [PubMed]
43. Qiu, Z.; Shu, J.; Liu, J.; Tang, D. Dual-channel photoelectrochemical ratiometric aptasensor with up-converting nanocrystals using spatial-resolved technique on homemade 3D printed device. *Anal. Chem.* **2019**, *91*, 1260–1268. [CrossRef] [PubMed]
44. Zhu, L.; Zhang, M.; Ye, J.; Yan, M.; Zhu, Q.; Huang, J.; Yang, X. Ratiometric electrochemiluminescent/electrochemical strategy for sensitive detection of microrna based on duplex-specific nuclease and multilayer circuit of catalytic hairpin assembly. *Anal. Chem.* **2020**, *92*, 8614–8622. [CrossRef] [PubMed]
45. Liu, Y.; Yang, H.; Wan, R.; Khan, M.; Wang, N.; Busquets, R.; Deng, R.; He, Q.; Zhao, Z. Ratiometric G-quadruplex assay for robust lead detection in food samples. *Biosensors* **2021**, *11*, 274. [CrossRef]

46. Zhang, G.; Chai, H.; Tian, M.; Zhu, S.; Qu, L.; Zhang, X. Zirconium-metalloporphyrin frameworks-luminol competitive electrochemiluminescence for ratiometric detection of polynucleotide kinase activity. *Anal. Chem.* **2020**, *92*, 7354–7362. [CrossRef]
47. Wang, L.; Xie, Y.; Yang, Y.; Liang, H.; Wang, L.; Song, Y. Electroactive covalent organic frameworks/carbon nanotubes composites for electrochemical sensing. *ACS Appl. Nano Mater.* **2020**, *3*, 1412–1419. [CrossRef]
48. Meng, Y.; Luo, Y.; Shi, J.; Ding, H.; Lang, X.; Chen, W.; Zheng, A.; Sun, J.; Wang, C. 2D and 3D porphyrinic covalent organic frameworks: The influence of dimensionality on functionality. *Angew. Chem. Int. Ed.* **2020**, *59*, 3624–3629. [CrossRef]
49. Liu, Y.; Ma, Y.; Zhao, Y.; Sun, X.; Gandara, F.; Furukawa, H.; Liu, Z.; Zhu, H.; Zhu, C.; Suenaga, K.; et al. Weaving of organic threads into a crystalline covalent organic framework. *Science* **2016**, *351*, 365–369. [CrossRef]
50. Liang, A.; Zhi, S.; Liu, Q.; Li, C.; Jiang, Z. A new covalent organic framework of dicyandiamide-benzaldehyde nanocatalytic amplification SERS/RRS aptamer assay for ultratrace oxytetracycline with the nanogold indicator reaction of polyethylene glycol 600 ACS. *Biosensor* **2021**, *11*, 458. [CrossRef]
51. Liang, H.; Wang, L.; Yang, Y.; Song, Y.; Wang, L. A novel biosensor based on multienzyme microcapsules constructed from covalent-organic framework. *Biosens. Bioelectron.* **2021**, *193*, 113553. [CrossRef] [PubMed]
52. Wu, N.; Wang, L.; Xie, Y.; Du, Y.; Song, Y.; Wang, L. Double signal ratiometric electrochemical riboflavin sensor based on macroporous carbon/electroactive thionine-contained covalent organic framework. *J. Colloid Interface Sci.* **2022**, *608*, 219–226. [CrossRef] [PubMed]
53. Dong, S.; Zhang, J.; Huang, G.; Wei, W.; Huang, T. Conducting microporous organic polymer with -OH functional groups: Special structure and multi-functional integrated property for organophosphorus biosensor. *Chem. Eng. J.* **2021**, *405*, 126682. [CrossRef]
54. Gutierrez-Sanchez, C.; Pita, M.; Vaz-Dominguez, C.; Shleev, S.; De Lacey, A. Gold nanoparticles as electronic bridges for laccase-based biocathodes. *J. Am. Chem. Soc.* **2012**, *134*, 17212–17220. [CrossRef]
55. Cai, C.X.; Ju, H.X.; Chen, H.Y. Electrocatalysis of poly-thionine modified microband gold electrode for oxidation of reduced nicotinamide adenine dinucleotide. *Chem. J. Chin. Univ.* **1995**, *16*, 368–372.
56. Liu, H.; Lu, X.; Xiao, D.; Zhou, M.; Xu, D.; Sun, L.; Song, Y. Hierarchical Cu–Co–Ni nanostructures electrodeposited on carbon nanofiber modified glassy carbon electrode: Application to glucose detection. *Anal. Methods* **2013**, *5*, 6360–6367. [CrossRef]
57. Huang, Y.; Zhang, Y.; Deng, X.; Li, J.; Huang, S.; Jin, X.; Zhu, X. Self-encapsulated enzyme through in-situ growth of polypyrrole for high-performance enzymatic biofuel cell. *Chem. Eng. J.* **2022**, *429*, 132148. [CrossRef]
58. Wang, Y.; Zhang, X.; Zhan, Y.; Li, J.; Nie, H.; Yang, Z. Au@Fe$_3$O$_4$ nanocomposites as conductive bridges coupled with a bi-enzyme-aided system to mediate gap-electrical signal transduction for homogeneous aptasensor mycotoxins detection. *Sens. Actuators B* **2020**, *321*, 128553. [CrossRef]
59. Zeng, R.; Huang, Z.; Wang, Y.; Tang, D. Enzyme-encapsulated DNA hydrogel for highly efficient electrochemical sensing glucose. *ChemElectroChem* **2020**, *7*, 1537–1541. [CrossRef]
60. Luo, Z.; Zhang, L.; Zeng, R.; Su, L.; Tang, D. Near-infrared light-excited core–core–shell UCNP@Au@CdS upconversion nanospheres for ultrasensitive photoelectrochemical enzyme immunoassay. *Anal. Chem.* **2018**, *90*, 9568–9575. [CrossRef]
61. Wang, X.; Yang, S.; Shan, J.; Bai, X. Novel electrochemical acetylcholinesterase biosensor based on core-shell covalent organic framework@multi-walled carbon nanotubes (COF@MWCNTs) composite for detection of malathion. *Int. J. Electrochem. Sci.* **2022**, *17*, 220543. [CrossRef]
62. Da Silva, M.; Vanzela, H.; Defavari, L.; Cesarino, I. Determination of carbamate pesticide in food using a biosensor based on reduced graphene oxide and acetylcholinesterase enzyme. *Sens. Actuators B* **2018**, *277*, 555–561. [CrossRef]
63. Santos, C.S.; Mossanha, R.; Pessoa, C.A. Biosensor for carbaryl based on gold modified with PAMAM-G$_4$ dendrimer. *J. Appl. Electrochem.* **2015**, *45*, 325–334. [CrossRef]
64. Gong, Z.; Guo, Y.; Sun, X.; Cao, Y.; Wang, X. Acetylcholinesterase biosensor for carbaryl detection based on interdigitated array microelectrodes. *Bioprocess Biosyst. Eng.* **2014**, *37*, 1929–1934. [CrossRef] [PubMed]
65. Liu, Q.; Fei, A.; Huan, J.; Mao, H.; Wang, K. Effective amperometric biosensor for carbaryl detection based on covalent immobilization acetylcholinesterase on multiwall carbon nanotubes/graphene oxide nanoribbons nanostructure. *J. Electroanal. Chem.* **2015**, *740*, 8–13. [CrossRef]
66. Sun, X.; Gong, Z.; Cao, Y.; Wang, X. Acetylcholinesterase biosensor based on poly (diallyldimethylammonium chloride)-multi-walled carbon nanotubes-graphene hybrid film. *Nano-Micro Lett.* **2013**, *5*, 47–56. [CrossRef]
67. Cesarino, I.; Moraes, F.; Lanza, M.; Machado, S. Electrochemical detection of carbamate pesticides in fruit and vegetables with a biosensor based on acetylcholinesterase immobilised on a composite of polyaniline–carbon nanotubes. *Food Chem.* **2012**, *135*, 873–879. [CrossRef]
68. Liu, B.; Xiao, B.; Cui, L. Electrochemical analysis of carbaryl in fruit samples on graphene oxide-ionic liquid composite modified electrode. *J. Food Compos. Anal.* **2015**, *40*, 14–18. [CrossRef]
69. Salih, F.; Achiou, B.; Ouammou, M.; Bennazha, J.; Ouarzane, A.; Alami Younssi, S.; Rhazi, M.E. Electrochemical sensor based on low silica X zeolite modified carbon paste for carbaryl determination. *J. Adv. Res.* **2017**, *8*, 669–676. [CrossRef]
70. Flavio, D.; Michele, D.; Manuel, S.; Dario, C.; Alberto, E. Press-transferred carbon black nanoparticles on board of microfluidic chips for rapid and sensitive amperometric determination of phenyl carbamate pesticides in environmental samples. *Microchim. Acta* **2016**, *183*, 3143–3149.
71. Zhang, M.; Zhang, Z.; Yang, Y.; Zhang, Y.; Wang, Y.; Chen, X. Ratiometric strategy for electrochemical sensing of carbaryl residue in water and vegetable samples. *Sensors* **2020**, *20*, 1524. [CrossRef] [PubMed]

Review

Progress in Probe-Based Sensing Techniques for In Vivo Diagnosis

Cheng Zhou [1,2], Zecai Lin [1,2], Shaoping Huang [1,3], Bing Li [4] and Anzhu Gao [1,2,*]

[1] Institute of Medical Robotics, Shanghai Jiao Tong University, Shanghai 200240, China
[2] Department of Automation, Shanghai Jiao Tong University, Shanghai 200240, China
[3] Department of Biomedical Engineering, Shanghai Jiao Tong University, Shanghai 200240, China
[4] Institute for Materials Discovery, University College London, London WC1E 7JE, UK
* Correspondence: anzhu_gao@sjtu.edu.cn

Abstract: Advancements in robotic surgery help to improve the endoluminal diagnosis and treatment with minimally invasive or non-invasive intervention in a precise and safe manner. Miniaturized probe-based sensors can be used to obtain information about endoluminal anatomy, and they can be integrated with medical robots to augment the convenience of robotic operations. The tremendous benefit of having this physiological information during the intervention has led to the development of a variety of in vivo sensing technologies over the past decades. In this paper, we review the probe-based sensing techniques for the in vivo physical and biochemical sensing in China in recent years, especially on in vivo force sensing, temperature sensing, optical coherence tomography/photoacoustic/ultrasound imaging, chemical sensing, and biomarker sensing.

Keywords: minimally invasive surgery; endoluminal intervention; physical sensing; biochemical sensing; in vivo diagnosis

1. Introduction

Surgical procedures can be divided into three main categories: open surgery, minimally invasive, and non-invasive surgery. Minimally invasive surgery (MIS), implementing the procedure through small incisions or natural orifices, has developed rapidly in recent decades, and continues to expand in many indications, such as cardiovascular, pancreatic, gastric, and endometrial cancers [1]. For example, the number of minimally invasive cardiovascular surgeries in 2019 in China reached 51,354, a 24.0% increase from 41,430 the year before [2]. Especially with the emergence of robotic surgery, minimally invasive surgery techniques relieve the patient's pain greatly, because of the shorter operation time, lower risk, small incision, and more precise medicine [3–5].

The structures of commercial needle [6], guidewire [7], catheter [8], balloon [9], continuum robots [10–13], and other medical instruments, whose outer diameter is down to several millimeters, submillimeters, or even micro-millimeters, are naturally fit for minimally invasive surgery due to their slender characteristics (typical lumens or cavities of humans are shown in Figure 1). Especially after introducing continuum robots to surgery, diagnosis or treatment can be introduced deep into the human anatomy after passing through tortuous lumens [14,15].

During the operation procedure, surgeons cannot recognize the precise status of the robots in an uncertain environment without the sensing capability. To improve surgical performance, medical robots or devices can be mounted with sensors to collect interaction or surrounding information, including contact force [16–18], surrounding temperature, vision [19,20], and geometrics or pathology inside the vessels [21–23]. Additionally, to detect diseases in their early stages before symptoms appear, it is significant to make deep tissue diagnoses. Biochemical sensing plays an important role in early diagnosis. When tissues produce adverse effects and pathological changes, the body release important

chemical signals or biomarkers, such as pH [24–26], superoxide anion [27], glucose [28–31], specific proteins [32], and H_2O_2 [33]. The expression of these signals often precedes the noticeable decline in tissue or organ function.

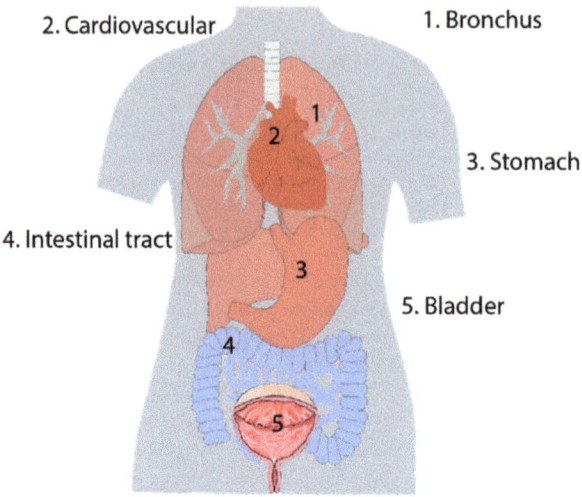

Figure 1. Typical lumens or cavities of humans.

In this paper, we focus on probe-based biosensors (as shown in Figure 2), including sensors with "probe" shapes, such as needle-based, guidewire-based, catheter-based, balloon-based, and continuum robot-based sensors. We review these sensors from the perspectives of in vivo physical sensing techniques, including force sensing, temperature sensing, optical coherence tomography (OCT) imaging, photoacoustic (PA) imaging, and ultrasound (US) imaging, and biochemical sensing techniques developed by domestic researchers in recent years. Based on this review, we aim to provide a different perspective focused on interventional surgery. This will support the optimization of the sensors' structural design and surface modification, and improve the devices' functionality and integration level.

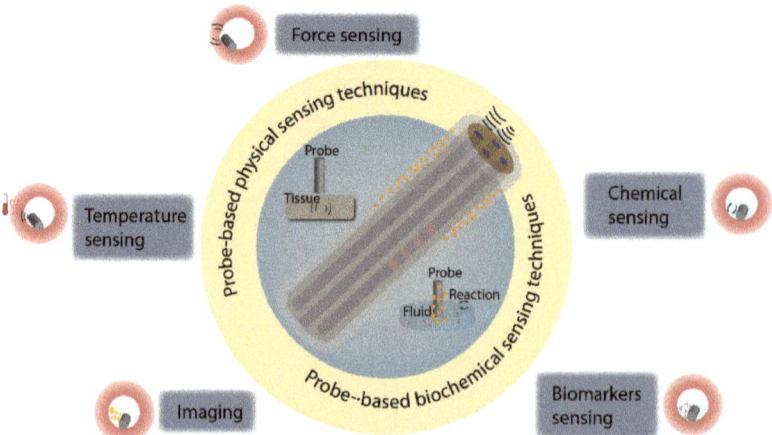

Figure 2. Overview of probe-based sensing techniques for in vivo diagnosis. Various probe-based sensors have been developed to detect physical and biochemical signals in vivo.

2. Probes for In Vivo Physical Sensing

2.1. Probes for In Vivo Force Sensing

Visual feedback is a basic function for surgeons in minimally invasive operations, but this single-mode feedback is not sufficient for delicate manipulation. Force sensing can provide intuitive, real-time, and interactive feedback to the operator. Therefore, it plays an important role in the high-quality diagnosis of biomedical environment [34,35].

2.1.1. Fiber Bragg Grating-Based Force Sensor

Fiber Bragg grating (FBG) is a periodic grating with a different refractive index to the core of silica optical fiber. The refractive index of periodic FBG is usually modified by UV exposition. The spatial period of the grating, as well as the refractive indices of the core, are quantities that depend on mechanical strain and temperature, resulting in the change of detected reflected wavelength. This can be used for accurate axial micro-strain measurements at high sampling rates in real time [36,37]. The principle of FBG for force sensing is shown in Figure 3.

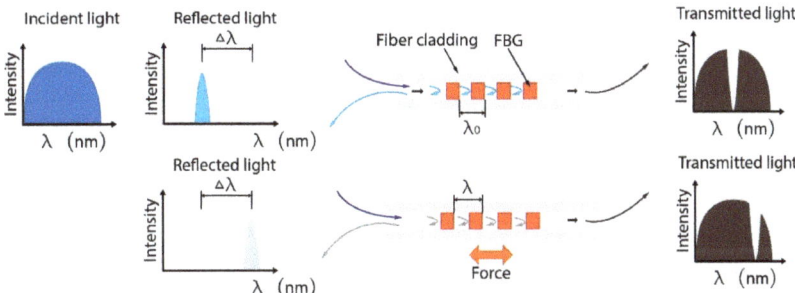

Figure 3. Mechanism of FBG-based force sensing. The FBG reflects the specific wavelength related to grating period, which can be changed with the axial strain. According to the physical characteristics and the reflected wavelength, the axis force can be calculated.

Optical fiber is a common optical waveguide for the transmission of light signals [38]. FBG-based optical fiber sensors have many advantages, such as high flexibility, lightweight, dielectric suitability, and MRI compatibility. Additionally, owing to optical fiber's inherent vantages that include miniaturized size, biocompatibility, and intrinsic sensing elements, it is suitable for fragile or confined environments, such as the intra-body parts of humans.

FBG sensors have been widely employed as surgical tools and biosensors, showing a great potential for biomedical engineering [39]. Flexible probes allow performing large-area tissue scanning or palpation for early stage cancer screening, which need to provide a gentle contact between the probe and tissue. During this procedure, a continuum manipulator equipped with a force/torque sensor is necessary [40]. Comparative studies between medical imaging-based visual operation and tactile force feedback-assisted operation reveal the superiority of interventional therapy with tactile force feedback [41]. The availability of tactile force feedback could reduce perioperative complications. Wang et al. proposed a scanning device for intraoperative thyroid gland endomicroscopy with one degree of freedom (DOF) FBG force sensor [42]. To obtain more information for clinical surgery, force-sensing devices with more degrees of freedom have been studied. Ping et al. presented a 3-DOF scanning device including an axial linear motion for approaching the tissue and two bending DOFs for surface scanning. To adapt the structure of commercial platforms, such as a standard endoscope, four FBG sensors were integrated into a 2.7 mm continuum robot. Three FBG sensors measured the transverse forces, while the other FBG sensor monitored the axial forces. Then, the flexible instrument with the FBG sensors was used for gastric endomicroscopy [43], as shown in Figure 4a. To avoid the failure of FBG, five single-mode fibers were buried into the deformable matrix in parallel rather than being simply pasted on the surface of robot. Ex vivo tissue palpation was implemented to validate the effectiveness of the sensor, and surface reaction forces and hard inclusions could be identified. In addition to producing force feedback for localizing tissue hard inclusions, an FBG-based fiber matrix was employed to reconstruct the surface profile of tissues during the process of palpation [44], as shown in Figure 4b. The moment can also influence the accuracy of sensing. The use of flexures coupled with FBG sensors has been demonstrated with high accuracy and repeatability for tissue force sensing. So, in Gao's work, a decoupling sensitivity matrix based on beam theory was presented to analyze the tip force and moment [45], as shown in Figure 4c.

In addition to tip force for palpation, lateral contact between the continuum robot body and the surrounding environment is unavoidable. The calculation of the distributed strain along the fiber body may provide diagnoses of diseases, such as motility of the gastrointestinal tract. Zhang et al. proposed a distributed hyperelastic elastomer-packaged pressure sensor for lateral force sensing [46].

However, temperature noise is a common negative interference for fiber-optical sensing. To reduce the thermal noise effect on biosensors, Ran et al. employed a single microfiber Bragg grating-based biosensor with second and third harmonic resonance. During the heating process, the third harmonic resonance held a distinct response with respect to the second harmonic resonance, and thus the thermal noise can be decoupled [47].

2.1.2. Electrical-Based Force Sensing

Electrical-based force sensing has the longest history of development and the most widespread application. A variety of electrical-based force sensors based on triboelectric nanogenerator, capacitive sensors, piezoresistive sensors, strain gauges, etc., have been attempted for minimally invasive surgical instruments. The capacitive theory is one of the most common methods used for force sensing. Senthil et al. proposed a stretchable capacitive-based pressure sensor patch, which can be integrated onto balloons towards continuous intra-abdominal pressure monitoring [48], for example, as shown in Figure 4d. Triboelectric nanogenerator, as a kind of new developing technology, and its application in force sensing has attracted more and more attention. Liu et al. reported an endocardial pressure sensor based on a triboelectric nanogenerator, which is not only flexible but also self-powered [8]. The sensor was assembled into a surgical catheter for minimally invasive surgery (MIS), and the endocardial pressure was monitored by using the changes in voltage. To sense the grasping force of surgical robots, some efforts need to be made at the integration of electrical-based sensors and small grippers. Hou et al. developed a biocompatible piezoresistive triaxial force sensor chip, which was integrated into the grip of a continuum robot to sense the grasping force of the MIS [49]. Yu et al. presented surgical forceps that consisted of double grippers with 3D pulling and grasp force sensing, and a simple structure with a double E-type strain beam was used as the substrate to minimize the size of the robot [50,51]. The main principle of these force sensing devices was to transform the voltage changes in the strain gauges into the force with a mathematical model. Machine learning is also used in electrical-based force sensing. Shi et al. proposed a method of force detection for a surgical robot, where the 3D force of the end-effector was decomposed by using an elastic element with an orthogonal beam structure. Moreover, a machine algorithm was used to learn the relationship between the acting force and the output voltage [52]. Other researchers also achieved the force sensing of robotics MIS with electromagnetic sensors, but most of the robots integrated with these sensors were hard to miniaturize, due to the rigid body or large diameter of the sensors.

2.1.3. Other Techniques

Two-photon printing is a powerful technique to fabricate micro- or nanostructures for force sensing. With the continuous miniaturization of surgical instruments, force sensors based on micro or nanostructures have gradually become a research hotspot. Zou et al. proposed a Fabry–Perot-based nanonewton-scale force sensor, which was printed on a single-mode fiber tip to measure the adhesion forces applied on the surfaces of micro/nanoscale structures [53]. A combination of multiple manufacturing methods can provide a possible for complex microstructure to sensing. Li et al. exploited micro-3D fabrication technology combining two photon polymerization and carbon-nanotube spraying techniques to construct a microspring-based electrical resistive sensor on the tip of a continuum robot. To demonstrate its potential, the device was employed to monitor human arterial pulses and real-time non-invasive intraluminal intervention [54], as shown in Figure 4e.

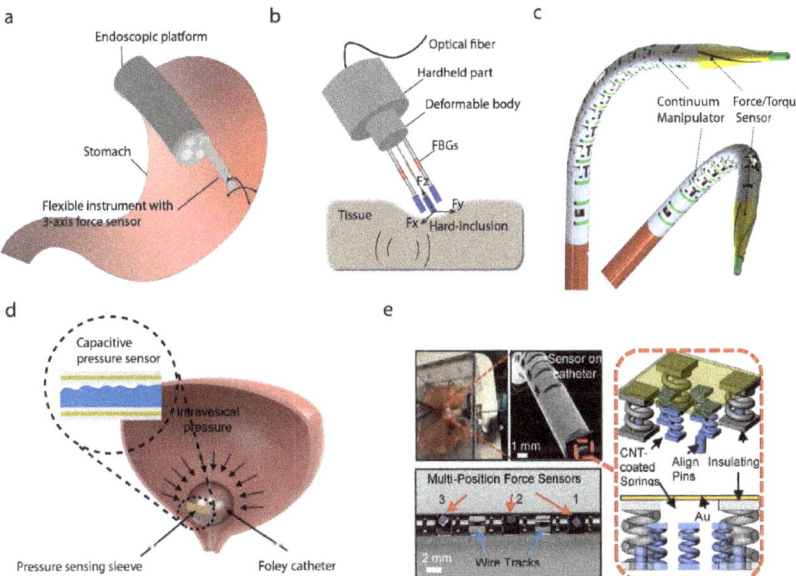

Figure 4. Probes for in vivo force sensing. (**a**) Flexible medical instrument for 3-axis force sensing [43]. (**b**) An elastic element with an orthogonal beam structure for end-effector 3D force decomposing [44]. (**c**) A spiral FBG force sensors-based method to measure the force and torque applied at the tip of the probe. Reprinted with permission from Ref. [45]. Copyright 2020, Elsevier. (**d**) A stretchable capacitive-based pressure sensor for continuous intra-abdominal pressure monitoring [48]. (**e**) Printed micro-spring with a carbon-nanotube force sensing layer on the tip of a continuum robot achieving a non-invasive intraluminal intervention. Reprinted with permission from Ref. [54]. Copyright 2019, American Chemical Society.

2.2. Probes for In Vivo Temperature Sensing

Thermal therapy, such as laser ablation and optogenetics, is a commonly surgical procedure in precision medicine. Ablation, especially laser treatment, is a medical procedure to remove diseased tissue and optogenetics is a method to study the relationship between biological conduction and light. To prevent the tissue from overheating during the process of the aforementioned operation, tissue temperature needs to be accurately monitored to ensure the quality and efficiency of therapy.

In order to achieve this, a flexible device, which contains platinum microsensors with a linear temperature–resistance relationship and flexible interconnects, was attached to the surface of the medical cryoballoon to detect the temperature distribution. Platinum (Pt) was chosen due to its properties of biocompatibility and thermal-resistance linearity. The temperature sensing ability of the device was verified by an ex vivo porcine heart cryoablation experiment [55], as shown in Figure 5a. Additionally, a silicon-based probe with Pt-based thin-film thermal-resistance sensor was proposed to reach deep tissues [56], as shown in Figure 5b.

Franz et al. reported the use of blackbody radiation in the short-wave infrared range for the tissue temperature monitoring during the laser vaporization. The devices integrated with the catheter to allow temperature sensing in vivo [57]. Ding et al. presented the optoelectronic devices to achieve an efficient NIR-to-visible upconversion for thermal detection, featuring high sensitivities and low-power excitation. Furthermore, the thermal sensors can be assembled as arrays to map spatial temperature, and an integrated optical fiber-thermometer device was employed for monitoring temperature variations in the deep brain of mice [58]. An optical fiber probe composed of FBG and a graphene oxide film coated S-shape fiber taper was made into a reflection type. The temperature measurements were realized by monitoring the characteristic wavelength shifts of the SFT and FBG in the reflection spectrum [59], as shown in Figure 5c. The master distributions of temperature in cardiac tissue during and after ablation an important to understand and implement this process. Koh et al. proposed an ultrathin and flexible needle-type system that could be inserted into the myocardial tissue in a minimally invasive way. It can be used to monitor the temperature in the transmural direction during the process of ablation. The measurement results exhibited excellent performance [60], as shown in Figure 5d.

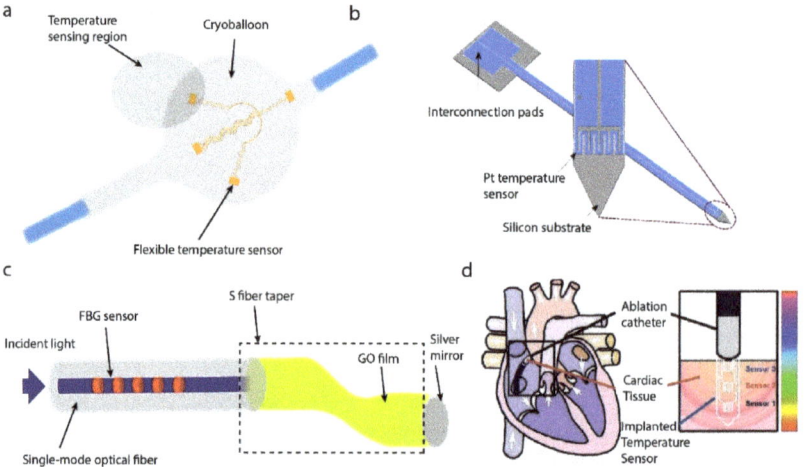

Figure 5. Probes for in vivo temperature sensing. (**a**) Flexible platinum-based temperature-sensing microsensors for cryoablation temperature monitoring [55]. (**b**) Pt-thermo-resistance-based temperature sensor for temperature monitoring to prevent overheating issues in optogenetics [56]. (**c**) Reflective FBG-based probe for temperature sensing [59]. (**d**) Ultrathin temperature sensor for cardiac ablation monitoring. Reprinted with permission from Ref. [60]. Copyright 2015, John Wiley and Sons.

2.3. Probes for In Vivo Imaging

2.3.1. Optical Coherence Tomography Imaging

OCT is a non-invasive three-dimensional imaging technique based on echo technology. It uses low-coherence light to scan tissues and capture the optical backscatter from deep tissues. It can provide cross-sectional images below the tissue surface rather than the outer surface images provided by microscopy. Due to its non-invasive, deep penetration, and high-resolution characteristics, the OCT technique has been widely used for medical imaging [61–63]. Compared with other imaging techniques such as magnetic resonance imaging (MRI) and optical microscopy, OCT possesses a higher resolution. The resolution can be classified into axial and transverse resolutions. The axial resolution can be expressed by $\Delta z = \frac{2\ln 2}{\pi} \frac{\lambda^2}{\Delta \lambda}$, determined by the center wavelength λ and the spectral bandwidth $\Delta \lambda$. The transverse resolution can be expressed by $\Delta x = \frac{4\lambda}{\pi} \frac{f}{d}$, determined by the center wavelength λ, the effective focal length f of the focusing optics, and the spot size d on the objective lens [64].

An endoscope integrated with OCT, such as intravascular optical coherence tomography (IVOCT), is a significant clinical application for in vivo imaging in situ. It enables obtaining information from a constrained space, such as blood vessels. These spaces are hard to image using traditional methods, such as optical microscopy. The IVOCT is commonly composed of a light source, a light detector, and a gradient-index lens with a microprism at the distal end. The light is emitted from the light source, and then the distal lens directs the light beam, resulting in a focused output beam perpendicular to the catheter axis [65,66].

To better understand physical information of the blood vascular, such as blood pressure, blood flow, and vascular stenosis, Wang et al. combined an OCT catheter with FBG to acquire shape parameters and OCT images in real time to reconstruct the vascular model [67]. Kang et al. demonstrated an all-fiber-based proximal-driven OCT catheter, in which the lens can directly be assembled with the commercial single-mode fiber [68]. The in vivo imaging capability of the catheter was evaluated in animal models. Li et al. presented a tri-modality intravascular imaging system that integrated a tri-modality probe, including OCT, ultrasound, and fluorescence imaging, which makes it promising for clinical management [69], as shown in Figure 6a.

2.3.2. Ultrasound Imaging

Ultrasound imaging, which is a safe, effective, and inexpensive technique for continuous monitoring in vivo, has been widely used in biomedical applications, particularly in clinical diagnosis, including extracranial steno-occlusive lesions diagnosis, intracranial stenosis diagnosis, and acute intracranial occlusion [70–73]. Especially in the process of stent graft placement, the use of intra-arterial is necessary. With the advancements in endovascular treatment and device fusion, the use of intravascular ultrasound incorporated with other techniques can treat more complex vascular pathologies [74,75]. Catheter-based intravascular ultrasound is a commonly used method to obtain real-time sufficient geometrical and pathological information from inside vessels for auxiliary diagnosis of intravascular diseases [76]. There are two typical catheter-based transducers: a mechanically rotating single-element transducer and an electronically phased array transducer. Zhang et al. focused on developing all-optical ultrasound probe for a miniature imaging system. They presented an ultrasound generator based on a step-indexed multimode fiber coated with a carbon nanotube and silicone rubber. The ultrasound was excited at the composite membrane with a nanosecond pulsed laser. Additionally, the ultrasound detector was made with rare-earth-doped fiber incorporating two reflective Bragg reflectors. The ultrasound probe was evaluated by imaging the cross-sectional structure of a swine trachea ex vivo [77], as shown in Figure 6b. To measure multiple parameters in arteries, Hong et al. designed a dual-mode ultrasound imaging catheter, including forwarding-looking and side-looking transducer. The transducers were composed of a piezoelectric layer, a top matching layer, and a backing layer. Tissue phantom experiments indicated that the dual-mode catheter could be possibly used for a one-time acquisition of multiple parameters, such as morphological and functional flow information about the vessel [78]. For calculating fractional flow reserve, a method combining coronary (X-ray) angiography curvature images with the intravascular ultrasound cross-section images was used to reconstruct the 3D structure. Based on the constructed structure, hemodynamics analysis was used to calculate fractional flow reserve. Aiming to obtain compact driving system, Wang et al. utilized miniature rotary–linear ultrasonic motor to drive the imaging catheter without any transmission structures. The motor could realize rotational and linear movements by changing ultrasonic motor modes. This work holds great promise for further compact system design [79].

2.3.3. Photoacoustic Imaging

Photoacoustic imaging (PAI), which makes high-resolution deep tissue visualization possible, is a fast-growing technology for medical diagnosis, such as cancer imaging [80,81]. The tissue absorbs laser energy instantaneously, when irradiated by pulse laser, and expands to produce ultrasonic waves. The ultrasonic signal is then received by sensors for post-processing. Photoacoustic imaging includes tomography and microscopic imaging. The former uses diffused light to stimulate tissues, resulting in an imaging depth of several centimeters and a spatial resolution of tens of microns. The latter uses a focused laser to irradiate biological tissue and detects ultrasonic signals, resulting in transverse resolution up to the submicron level [82].

Fiber integrated with a PAI device was used to image biological tissue at a subcellular resolution in a minimally invasive manner [83]. Photoacoustic imaging microscopy (PAM) excites a focused pulsed laser beam and detects ultrasound waves through a piezoelectric transducer, and then generates images in a line. Moving the scanner could obtain two- or three-dimensional imaging [84]. In order to image in a constrained environment, steering imaging transducers will offer more applications. An ultrasound beam based on photoacoustic effect was excited by laser irradiating the gold nanocomposites, which was dip-coated on the tip of an optical fiber, and then a micro concave prism was fixed on the gold nanocomposites' surface to focus the ultrasound waves. The ultrasound echo was received by a Fabry–Perot fiber optic sensor to reconstruct the ultrasound field. The potential of this minimally invasive diagnostics was implemented on an ex vivo porcine tissue experiment [85]. To conveniently identify the area of interest, miniature endoscopy combining photoacoustic microscopy and white-light microscopy was proposed. The white-light microscopy can guide the PAM to the region of interest before imaging [86], as shown in Figure 6c.

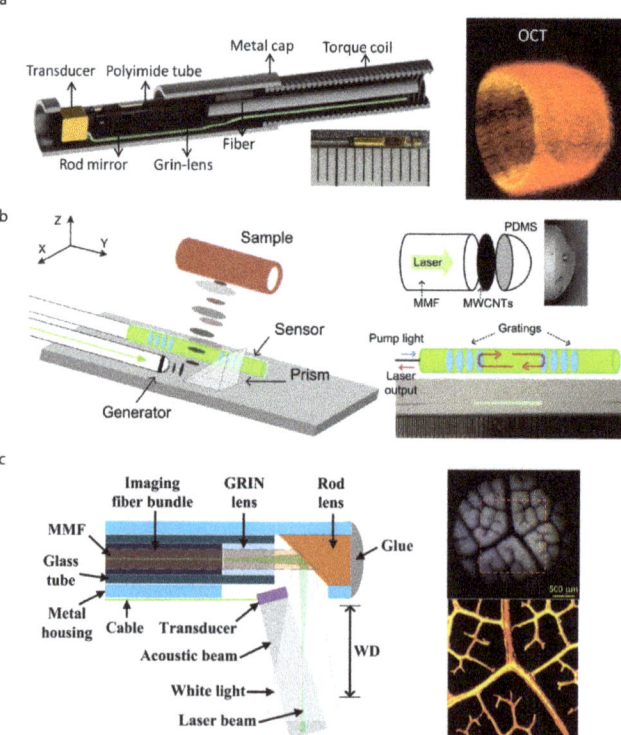

Figure 6. Probes for in vivo imaging. (**a**) Tri-modality probe and OCT image [69]. (**b**) The 125 um all-fiber-based ultrasound probe [77]. (**c**) Miniature probe-based photoacoustic microscopy and white-light microscopy endoscope. Reprinted with permission from Ref. [86]. Copyright 2020, John Wiley and Sons.

Catheter-based intravascular photoacoustic imaging is an emerging modality. It possesses many advantages. For example, the imaging depth of IVPA has been extended beyond the ballistic regime and IVPA can share the same detector with IVUS imaging, resulting in more complementary information of the tissue [87]. A miniature full field-of-view photoacoustic/ultrasonic endoscopic catheter system was used to depict the vasculature and morphology of the GI tract in vivo [88]. Cao et al. arranged a multimode fiber and acoustic device in a catheter tip to obtain efficient imaging overlap. The imaging capability was evaluated in a diseased porcine carotid artery and a human coronary artery ex vivo [87].

2.3.4. Other Techniques

Raman spectroscopy is a vibrational spectroscopy capable of probing biomolecular information, which has wide applications in cell/tissue characterization and diagnosis without labeling. The development of the fiber-based Raman endoscopic probe makes imaging internal organs possible, such as diagnosing nasopharyngeal carcinoma [89]. An outer diameter 300-micron optical fiber with semispherical ball lens tip was inserted into a syringe needle for a deep-tissue Raman imaging [90].

The diagnostic value of probe-based confocal laser endomicroscopy (pCLE) has been recognized in many medical fields, such as endoscopic surveillance of Barrett's esophagus [91], diagnosis of gastric carcinoma and precancerous lesions [92], and colon polyp histology [93]. Table 1 shows the comparison between these main in vivo imaging techniques [94].

Table 1. Comparison between in vivo imaging techniques.

Imaging Techniques	Axial Resolution	Transverse Resolution	Penetrating Depth	Integrated Size	Typical Applications
OCT imaging	0.2–1 μm [63]	0.6–2 μm [63]	1–2 mm [76]	Submillimeter [65,67,68]	Vascular shape reconstruction [67] Intracoronary optical coherence tomography [68]
US imaging	20–200 μm [72]	120–250 μm [76]	7–15 mm [76]	Submillimeter [77] Millimeter [78,79]	Intravascular imaging [74,78,79] Trachea imaging [77]
PA imaging	From sub-micrometer to sub-millimeter [94]	From sub-micrometer to sub-millimeter [94]	From sub-millimeter to depths up to several millimeters [94]	Submillimeter [85] Millimeter [73,75,86–88]	Tissue imaging [85,86,88] Intravascular imaging [87]

3. Probes for In Vivo Biochemical Sensing

3.1. Probes for In Vivo Chemical Sensing

A chemical sensing system contains receptors for the target biomolecules in the tissue fluid, and the transducers to convert the results into measurable signals, such as electrical or optical signals [95,96]. Cancer is a common and intractable disease all over the world. There are many efforts devoted to defeating cancer. A rapid and effective diagnosis is significant. In the comparison of normal differentiated adult and cancer cells, the extracellular pH of the former is about 7.4, but cancer cells have a higher pH. Dysregulated pH is a well-known cancer indicator, which could be employed to diagnose cancer [97,98]. Optical fiber-based pH sensors have the ability to measure pH in deep tissues. Chen et al. proposed a miniaturized probe to detect the pH of the cancer tissue environment, based on the ratio fluorescence method. The 520 nm laser passed through the fiber and irradiated the fluorophore, which changed with the pH and the spectral band remained almost unchanged.

The fluorescence emission light came back from the inner cladding, and then the returned light wave was detected by a spectrometer [24]. A U-shaped multimode optical fiber was demonstrated by Tang et al. The U-shaped bare region was coated with a hybrid organic–inorganic composite film, as a pH-sensitive layer. This bonding was affected by the hydrogen concentration in the solution, resulting in a refractive index change in the film. The sensor can be used to monitor the pH of human serum [25], as shown in Figure 7a. A hydrogel-based fiber pH sensor was demonstrated by Gong et al. The sensor was in situ photo-polymerized on the optical fiber tip with pH-sensitive hydrogel, the principle of which is the same as the fluorophore mentioned before. This optical fiber sensor was used to measure the pH of cancerous lung tissue [26], as shown in Figure 7b. To build fully biocompatible systems, Li et al. constructed in vivo red blood cells waveguide using two fiber probes in a microfluidic capillary to construct biosensors. By detecting the light propagation mode of the biosensor, the pH of blood can be detected in real time with high accuracy [99], as shown in Figure 7c. Peng et al. introduced a carbon-fiber microelectrode-based chemical sensor for monitoring the superoxide anion. The superoxide anion was acknowledged to be related to the development of many neurological diseases, including Alzheimer's disease [27], as shown in Figure 7d.

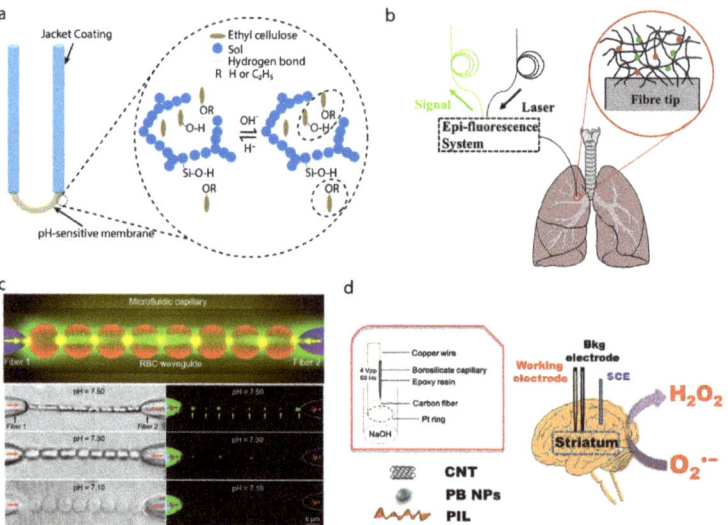

Figure 7. Probes for in vivo chemical sensing. (**a**) A U-shaped optical pH sensor based on hydrogen bonding [25]. (**b**) A hydrogel-based optical-fiber fluorescent pH sensor. Reprinted with permission from Ref. [26]. Copyright 2020, Elsevier. (**c**) Red-blood-cells waveguide for blood pH detection in real time. Reprinted with permission from Ref. [99]. Copyright 2019, John Wiley and Sons. (**d**) In vivo monitoring of superoxide anion with functionalized ionic liquid polymer-decorated microsensor. Reprinted with permission from Ref. [27]. Copyright 2019, Elsevier.

3.2. Probes for In Vivo Biomarkers Sensing

Blood glucose concentration is a typical parameter to represent the human metabolic level. Chen et al. coated needle electrodes with polyaniline nanofiber, platinum nanoparticles, glucose oxidase enzyme, and porous layers with a layer-by-layer deposition process [28]. Nanoparticles incorporated into conductive polyaniline nanofibers resulted in a high surface-to-volume ratio for the immobilization of electrocatalytic glucose enzyme. The performance was then tested by inserting the needle into mice models, showing an excellent response to the concentration of blood glucose and good biocompatibility with the tissue. A minimally invasive glucose probe with an electropolymerized conductive polymer polyaniline core capable of continuously monitoring subcutaneous glucose has been developed [29]. In vivo experiments using mice models showed the real-time response to the variation of blood glucose, as shown in Figure 8a. Additionally, a chitosan/sodium-alginate-modified polysulfone hollow fibrous membrane was fixed on the stainless-steel needle electrode [30]. The needle electrode can be inserted into the skin to record responsive currents to detect blood glucose, as shown in Figure 8b. With the help of advanced manufacturing techniques, a Fabry–Perot cavity sensor was printed on the tip of a single-mode optical fiber by two-photon printing for glucose detection. The sensor was sensitive to the refractive index changes induced by the concentration changes in glucose [31], as shown in Figure 8c.

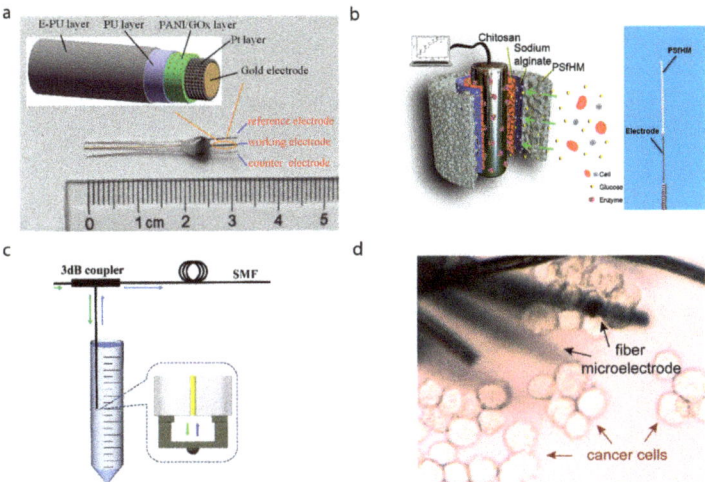

Figure 8. Probes for in vivo biomarkers sensing. (**a**) A needle-type blood glucose biosensor for long-term in vivo monitoring. Reprinted with permission from Ref. [29]. Copyright 2017, Elsevier. (**b**) Implantable glucose biosensing probe. Reprinted with permission from Ref. [30]. Copyright 2019, American Chemical. Society. (**c**) Fiber-tip micro Fabry–Perot interferometer for glucose concentration measurement. The Fabry–Perot interferometer is sensitive to the refractive index changes in analytes [31]. (**d**) Flexible nanohybrid microelectrode arrays for in situ biomarker H_2O_2 detection in live cancer cells. Reprinted with permission from Ref. [33]. Copyright 2018, Elsevier.

Wang et al. demonstrated a new type of functionalized multi-walled carbon nanotubes twisted fiber bundles to monitor multiple disease biomarkers, such as ions and prostate-specific antigens, hydrogen peroxide, and glucose [100]. Zhang et al. developed a flexible microelectrode based on carbon fiber wrapped by gold-nanoparticle-decorated nitrogen-doped carbon nanotube arrays, and researched its clinical applications in detecting the biomarker H_2O_2 expressed by living cancer cells in situ [33], as shown in Figure 8d.

Surface plasmon resonance (SPR) possesses a high compatibility with fiber-optic techniques. The sensors are sensitive to the refractive index of the certain materials. SPR sensors are usually constructed by coating a metal film on fiber surfaces, where reactions occur between the film and the environment. Additionally, the reaction changes the complex refractive index of the medium near the sensor surface, and therefore the SPR condition. Integrating SPR devices with an optical fiber can empower the biosensing systems with easily reading and in vivo monitoring capability [101]. Guo et al. presented a biosensor by coating a nanometer-scale silver film on tilted fiber Bragg grating to detect urinary protein variations. The sensors have the potential for a narrow endoluminal intervention in vivo [101]. A rare-earth-modified photothermal FBG-based fiber with fiber-optic fluorescent sensor was proposed to detect tumors in vivo. The fiber probe can turn the 450 nm excitation laser to an echo wavelength of 550 nm under a hypoxia tumor microenvironment and then kill the tumor through the photothermal effect [102].

4. Conclusions

This review focused on probe-based sensing techniques for the endoluminal intervention of minimally invasive or non-invasive procedures. It aimed to help researchers in the field of in vivo sensing techniques to diagnose/treat diseases, manipulate/assist interventional medical tools, and understand the latest domestic progress in recent years. The structure of the probe with a miniature size naturally fits minimally invasive procedures and enables a deeper tissue operation. A large amount of physiological information leads to a variety of in vivo sensing technologies, including physical sensing and biochemical sensing, in recent years. Physical sensing is the most direct perception for surgical intervention. For force sensing, in order to better control surgical instruments and improve surgical safety, the force feedback between instruments and tissues is important for both surgeons and patients. For temperature sensing, the real-time monitoring of the tissue temperature can improve the effectiveness of ablation. Research on the pathological characteristics of vascular diseases and the morphology of vessels has attracted vast attention from clinical medical researchers. Optical coherence tomography imaging, ultrasound imaging, photoacoustic imaging, and other imaging techniques can penetrate tissues, such as the vessel wall, to image its deep morphology to provide precise diagnosis. Different from these physical sensing techniques, biochemical sensing plays a more important role in early diagnosis rather than surgical operation. When tissues produce adverse effects and pathological changes in the initial stage of a disease, some important chemicals or biomarkers, such as pH, glucose, protein, and H_2O_2, often precedes a detectable decline in function. Thus, biochemical sensing can reflect the development of the disease, but physical sensing cannot distinguish the differences very well. Table 2 shows the comparison between these main probe-based sensing techniques.

Table 2. Comparison between probe-based sensing techniques.

Sensing Techniques	Sensing Mechanism	Carrier	Overall Diameter	Medical Scenario
Force sensing	Fiber Bragg grating sensors [42–46] Plastic Fiber Bragg grating sensors [36] Triboelectric nanogenerator [8] Piezoresistive [49] Piezoelectric Strain gauges [50–52] Capacitive pressure sensor [48] Closed-loop force control[37] Fiber-tip microforce sensor [53] Carbon nanotube-coated microsprings [54]	Continuum robot [43,45,49,54] Scanning device [42] Probe [44] Polymer package [46] Plastic optical fiber [36] Catheter [8] Gripper [50–52] Foley catheter balloon [48] Optical fiber [53]	Submillimeter [36,48,53] Millimeter [42–46,49–51,54] Centimeter [8,37,52]	Thyroidectomy [42] Gastric Endomicroscopy [43] Hard-inclusion Localization [44] Optical biopsy [45] In vivo pressure sensor [46] Blood pressure [36] Confocal laser endomicroscopy [37] Endocardial pressure monitoring [8] Three-dimensional force sensing for forceps [49–52] Intra-abdominal pressure monitoring [48] Measurement of interfacial adhesion force [53] Transcutaneous monitoring of human arterial pulses [54]

Table 2. Cont.

Sensing Techniques	Sensing Mechanism	Carrier	Overall Diameter	Medical Scenario
Temperature sensing	Thermo-resistance effect[55,56,60] Infrared-to-visible upconversion [58] Thermal expansion and thermal-optic effects [59] Short-wave infrared [57]	Balloon [55] Silicon-based probe [56] Needle-type polymer [60] Silica fiber [58] Optical fiber [59] Silica fiber [57]	Submillimeter [56–59] Millimeter [55,60]	Cryoablation [55] Optogenetic [56] Arrhythmias [60] Deep-brain thermal detection [58] Laser vaporization [57]
Optical coherence tomography imaging	Light scattering [62,67–69]	Catheter [62,67–69]	Submillimeter [67,68] Millimeter [69]	Intravascular Imaging [62,67–69]
Ultrasound imaging	Pulse-echo [74,77–79]	Catheter [74,78,79] Optical fiber [77]	Submillimeter [79] Millimeter [78]	Intravascular Imaging [74,78,79] Trachea imaging [77]
Photoacoustic imaging	Pulse-echo [85–88]	Catheter [86,88] Optical fiber [85]	Millimeter [86–88]	Tissue imaging [85,86,88] Intravascular imaging [87]
Chemical sensing	pH sensitivity of fluorophore [24,26] Polymer aggregation leads to refractive index changes [25] Enzymatic catalysis and electrochemical reactions [27]	Optical fiber [24,26] Organic–inorganic composite film-coated optical fiber [25] Carbon fiber microelectrode [27]	Submillimeter [24–26]	Discrimination of tumorous and normal tissues [24,26] Chronic wounds and/or fetal acidosis [25] Superoxide anion detection [27]
Biomarker sensing	Enzymatic catalysis and electrochemical reactions [28–30,33] Fabry–Perot (FP) cavity biosensor [31] Surface plasmon resonance [32]	Implantable electrode [28–30,33] Optical fiber [31,32]	0.005–0.03 mm [33] Submillimeter [28,31,32] Millimeter [29,30]	Blood glucose detection [28–31] Urinary protein detection [32] Cancer biomarker H_2O_2 detection [33]

There are many kinds of force sensors, including but not limited to FBG-based sensors, triboelectric nanogenerators, capacitive sensors, piezoresistive sensors, and strain gauges. Some of them are stable, reliable, and easy to miniaturize, but have large modulus and single measurement modes, such as FBG-based sensors. Some of them are soft, sensitivity-adjustable, and self-powered, but their applications may be compromised by their unstable performance, difficulty of miniaturizing, and the difficulty to producing plentifully. As for in vivo temperature sensing, it can be used to monitor thermal effects to avoid overheating, but it has not been widely used in the clinical scenario and can easily be replaced by other technologies, such as visual-based sensors. For in vivo imaging, IVUS is the most commonly used method in intravascular imaging, and OCT possesses a high resolution of intravascular imaging and has been applied to ophthalmology and other surgeries. The US has a more extensive imaging range and can penetrate tissues 7–15 mm deeper, while the OCT only has a 1–2 mm penetration power. Considering the high resolution of light-based OCT imaging techniques, the OCT can easily be disturbed by the environment, such as the scattering process in the background. Thus, the combination of US and OCT imaging techniques can achieve macroscopic and microscopic pathological characteristics and morphology of the tissue. PA imaging combines the high-contrast characteristics of optical imaging and the high-penetration-depth characteristics of ultrasound imaging, which can provide high-resolution and high-contrast tissue imaging. PA's imaging depth and spatial resolution in tissues are related to acoustic frequency [94]. For in vivo biochemical sensing, Doctors may obtain and evaluate symptoms in situ via markers most related to it, rather than judging the symptoms through indirect features that may lead to misjudgments. However, the lack of standard micro/nano fabrication approaches, long-term stability, and repeatability are still challenging problems for these sensors, especially when they are used in confined anatomical scenarios.

These probe-based sensing techniques for in vivo diagnosis do help the doctor to understand the patients' conditions more quickly and directly. However, even though these probe-based in vivo sensing techniques are implemented in minimally invasive way, these operation procedures are still not easy to conduct. These problems further hinder the commercialization of probe-based sensors. However, through successive efforts, in vivo sensing is becoming more precise, more instantaneous, and easier.

Author Contributions: Writing—review and editing, A.G., C.Z., Z.L., S.H. and B.L.; supervision, A.G. All authors have read and agreed to the published version of the manuscript.

Funding: This work was supported by the National Natural Science Foundation of China (62003209); the Natural Science Foundation of Shanghai (21ZR1429500); the Shanghai Rising Star Program (22QC1401400); the Science and Technology Commission of Shanghai Municipality (21511102100, 21DZ1100300, 20DZ2220400); SJTU Global Strategic Partnership Fund (2021 SJTU-UCL); the State Key Laboratory of Robotics (2021-KF-22-09).

Informed Consent Statement: Not applicable.

Data Availability Statement: Not applicable.

Conflicts of Interest: The authors declare no conflict of interest.

References

1. Concin, N.; Matias-Guiu, X.; Vergote, I.; Cibula, D.; Mirza, M.R.; Marnitz, S.; Ledermann, J.; Bosse, T.; Chargari, C.; Fagotti, A.; et al. ESGO/ESTRO/ESP guidelines for the management of patients with endometrial carcinoma. *Int. J. Gynecol. Cancer* **2021**, *31*, 12–39. [CrossRef] [PubMed]
2. Wei, L.; Chen, N.; Yang, Y.; Zheng, Z.; Dong, N.; Guo, H.; Mei, J.; Xue, S.; Liu, L.; Guo, Y. The data of Chinese minimally invasive cardiovascular surgery in 2019. *Chin. J. Clin. Thorac. Cardiovasc. Surg.* **2021**, *12*, 149–153.
3. Fuchs, K. Minimally invasive surgery. *Endoscopy* **2002**, *34*, 154–159. [CrossRef]
4. Burgner-Kahrs, J.; Rucker, D.C.; Choset, H. Continuum Robots for Medical Applications: A Survey. *IEEE Trans. Robot.* **2015**, *31*, 1261–1280. [CrossRef]
5. Runciman, M.; Darzi, A.; Mylonas, G.P. Soft Robotics in Minimally Invasive Surgery. *Soft Robot.* **2019**, *6*, 423–443. [CrossRef]
6. Hou, B.; Wang, M.; Song, Z.; He, Q.; Hao, Z. Renal puncture access using a blunt needle: Proposal of the blunt puncture concept. *World J. Urol.* **2022**, *40*, 1035–1041. [CrossRef] [PubMed]
7. Shen, H.; Wang, C.; Xie, L.; Zhou, S.; Gu, L.; Xie, H. A novel remote-controlled robotic system for cerebrovascular inter-vention. *Int. J. Med. Robot. Comput. Assist. Surg.* **2018**, *14*, e1943. [CrossRef]
8. Liu, Z.; Ma, Y.; Ouyang, H.; Shi, B.; Li, N.; Jiang, D.; Xie, F.; Qu, D.; Zou, Y.; Huang, Y. Transcatheter self-powered ul-trasensitive endocardial pressure sensor. *Adv. Funct. Mater.* **2019**, *29*, 1807560. [CrossRef]
9. Han, M.; Chen, L.; Aras, K.; Liang, C.; Chen, X.; Zhao, H.; Li, K.; Faye, N.R.; Sun, B.; Kim, J.-H.; et al. Catheter-integrated soft multilayer electronic arrays for multiplexed sensing and actuation during cardiac surgery. *Nat. Biomed. Eng.* **2020**, *4*, 997–1009. [CrossRef]
10. Ai, X.; Gao, A.; Lin, Z.; He, C.; Chen, W. A Multi-Contact-Aided Continuum Manipulator with Anisotropic Shapes. *IEEE Robot. Autom. Lett.* **2021**, *6*, 4560–4567. [CrossRef]
11. Wang, L.; Guo, C.F.; Zhao, X. Magnetic soft continuum robots with contact forces. *Extreme Mech. Lett.* **2022**, *51*, 101604. [CrossRef]
12. Li, W.; Shen, M.; Gao, A.; Yang, G.-Z.; Lo, B. Towards a Snake-Like Flexible Robot for Endoscopic Submucosal Dissection. *IEEE Trans. Med Robot. Bionics* **2020**, *3*, 257–260. [CrossRef]
13. Han, Y.; Marvi, H.; Sitti, M. Fiberbot: A miniature crawling robot using a directional fibrillar pad. In Proceedings of the 2015 IEEE International Conference on Robotics and Automation (ICRA), Seattle, WA, USA, 26–30 May 2015.
14. Fu, Y.; Gao, A.; Liu, H.; Guo, S. The master-slave catheterisation system for positioning the steerable catheter. *Int. J. Mechatron. Autom.* **2011**, *1*, 143. [CrossRef]
15. Gao, A.; Liu, H.; Zou, Y.; Wang, Z.; Liang, M.; Wang, Z. A Contact-Aided Asymmetric Steerable Catheter for Atrial Fibrillation Ablation. *IEEE Robot. Autom. Lett.* **2017**, *2*, 1525–1531. [CrossRef]

16. Fontanelli, G.A.; Buonocore, L.R.; Ficuciello, F.; Villani, L.; Siciliano, B. An External Force Sensing System for Minimally Invasive Robotic Surgery. *IEEE/ASME Trans. Mechatron.* **2020**, *25*, 1543–1554. [CrossRef]
17. Gao, A.; Zhou, Y.; Cao, L.; Wang, Z.; Liu, H. Fiber Bragg Grating-Based Triaxial Force Sensor with Parallel Flexure Hinges. *IEEE Trans. Ind. Electron.* **2018**, *65*, 8215–8223. [CrossRef]
18. Heunis, C.M.; Belfiore, V.; Vendittelli, M.; Misra, S. Reconstructing Endovascular Catheter Interaction Forces in 3D using Multicore Optical Shape Sensors. In Proceedings of the 2019 IEEE/RSJ International Conference on Intelligent Robots and Systems (IROS), Macau, China, 3–8 November 2019.
19. Lin, Z.; Gao, A.; Ai, X.; Gao, H.; Fu, Y.; Chen, W.; Yang, G.-Z. ARei: Augmented-Reality-Assisted Touchless Teleoperated Robot for Endoluminal Intervention. *IEEE/ASME Trans. Mechatron.* **2021**, *27*, 3144–3154. [CrossRef]
20. Wu, B.; Wang, L.; Liu, X.; Wang, L.; Xu, K. Closed-loop pose control and automated suturing of continuum surgical ma-nipulators with customized wrist markers under stereo vision. *IEEE Robot. Autom. Lett.* **2021**, *6*, 7137–7144. [CrossRef]
21. Beaudette, K.; Li, J.; Lamarre, J.; Majeau, L.; Boudoux, C. Double-Clad Fiber-Based Multifunctional Biosensors and Mul-timodal Bioimaging Systems: Technology and Applications. *Biosensors* **2022**, *12*, 90. [CrossRef]
22. Miranda, B.; Rea, I.; Dardano, P.; De Stefano, L.; Forestiere, C. Recent Advances in the Fabrication and Functionalization of Flexible Optical Biosensors: Toward Smart Life-Sciences Applications. *Biosensors* **2021**, *11*, 107. [CrossRef]
23. Hamed, A.; Tang, S.C.; Ren, H.; Squires, A.; Payne, C.; Masamune, K.; Tang, G.; Mohammadpour, J.; Tse, Z.T.H. Advances in Haptics, Tactile Sensing, and Manipulation for Robot-Assisted Minimally Invasive Surgery, Noninvasive Surgery, and Diagnosis. *J. Robot.* **2012**, *2012*, 1–14. [CrossRef]
24. Chen, M.; Wang, J.; Tan, W.; Feng, Y.; Zheng, G. Miniaturized all fiber probe for optical coherence tomography and pH detection of biological tissue. *J. Biophotonics* **2020**, *14*, e202000239. [CrossRef] [PubMed]
25. Tang, Z.; Gomez, D.; He, C.; Korposh, S.; Morgan, S.P.; Correia, R.; Hayes-Gill, B.; Setchfield, K.; Liu, L. A U-shape fi-bre-optic pH sensor based on hydrogen bonding of ethyl cellulose with a sol-gel matrix. *J. Light. Technol.* **2021**, *39*, 1557–1564. [CrossRef]
26. Gong, J.; Tanner, M.G.; Venkateswaran, S.; Stone, J.M.; Zhang, Y.; Bradley, M. A hydrogel-based optical fibre fluorescent pH sensor for observing lung tumor tissue acidity. *Anal. Chim. Acta* **2020**, *1134*, 136–143. [CrossRef] [PubMed]
27. Peng, Q.; Yan, X.; Shi, X.; Ou, S.; Gu, H.; Yin, X.; Shi, G.; Yu, Y. In vivo monitoring of superoxide anion from Alzheimer's rat brains with functionalized ionic liquid polymer decorated microsensor. *Biosens. Bioelectron.* **2019**, *144*, 111665. [CrossRef] [PubMed]
28. Chen, D.; Wang, C.; Chen, W.; Chen, Y.; Zhang, J.X. PVDF-Nafion nanomembranes coated microneedles for in vivo transcutaneous implantable glucose sensing. *Biosens. Bioelectron.* **2015**, *74*, 1047–1052. [CrossRef] [PubMed]
29. Fang, L.; Liang, B.; Yang, G.; Hu, Y.; Zhu, Q.; Ye, X. A needle-type glucose biosensor based on PANI nanofibers and PU/E-PU membrane for long-term invasive continuous monitoring. *Biosens. Bioelectron.* **2017**, *97*, 196–202. [CrossRef]
30. Zhou, J.; Ma, Z.; Hong, X.; Wu, H.-M.; Ma, S.-Y.; Li, Y.; Chen, D.-J.; Yu, H.-Y.; Huang, X.-J. Top-down strategy of im-plantable biosensor using adaptable, porous hollow fibrous membrane. *ACS Sensors* **2019**, *1*, 931–937. [CrossRef]
31. Tang, S.; Zou, M.; Zhao, C.; Jiang, Y.; Chen, R.; Xu, Z.; Yang, C.; Wang, X.; Dong, B.; Wang, Y.; et al. Fabry-Perot Interferometer Based on a Fiber-Tip Fixed-Supported Bridge for Fast Glucose Concentration Measurement. *Biosensors* **2022**, *12*, 391. [CrossRef]
32. Guo, T. Fiber Grating-Assisted Surface Plasmon Resonance for Biochemical and Electrochemical Sensing. *J. Light. Technol.* **2017**, *35*, 3323–3333. [CrossRef]
33. Zhang, Y.; Xiao, J.; Sun, Y.; Wang, L.; Dong, X.; Ren, J.; He, W.; Xiao, F. Flexible nanohybrid microelectrode based on carbon fiber wrapped by gold nanoparticles decorated nitrogen doped carbon nanotube arrays: In situ electrochemical detection in live cancer cells. *Biosens. Bioelectron.* **2018**, *100*, 453–461. [CrossRef]
34. Al-Ahmad, O.; Ourak, M.; Vlekken, J.; Poorten, E.V. FBG-Based Estimation of External Forces along Flexible Instrument Bodies. *Front. Robot. AI* **2021**, *8*, 718033. [CrossRef]
35. Bandari, N.; Dargahi, J.; Packirisamy, M. Tactile Sensors for Minimally Invasive Surgery: A Review of the State-of-the-Art, Applications, and Perspectives. *IEEE Access* **2019**, *8*, 7682–7708. [CrossRef]
36. Haseda, Y.; Bonefacino, J.; Tam, H.-Y.; Chino, S.; Koyama, S.; Ishizawa, H. Measurement of Pulse Wave Signals and Blood Pressure by a Plastic Optical Fiber FBG Sensor. *Sensors* **2019**, *19*, 5088. [CrossRef]
37. Latt, W.T.; Newton, R.C.; Visentini-Scarzanella, M.; Payne, C.J.; Noonan, D.P.; Shang, J.; Yang, G.-Z. A Hand-held Instrument to Maintain Steady Tissue Contact during Probe-Based Confocal Laser Endomicroscopy. *IEEE Trans. Biomed. Eng.* **2011**, *58*, 2694–2703. [CrossRef]
38. Liu, Z.; Zhang, Z.; Tam, H.-Y.; Tao, X. Multifunctional Smart Optical Fibers: Materials, Fabrication, and Sensing Applications. *Photonics* **2019**, *6*, 48. [CrossRef]
39. Presti, D.L.; Massaroni, C.; Leitao, C.S.J.; Domingues, M.D.F.; Sypabekova, M.; Barrera, D.; Floris, I.; Massari, L.; Oddo, C.M.; Sales, S.; et al. Fiber Bragg Gratings for Medical Applications and Future Challenges: A Review. *IEEE Access* **2020**, *8*, 156863–156888. [CrossRef]
40. Wu, Z.; Gao, A.; Liu, N.; Jin, Z.; Yang, G.-Z. FBG-Based Triaxial Force Sensor Integrated with an Eccentrically Configured Imaging Probe for Endoluminal Optical Biopsy. In Proceedings of the 2020 IEEE International Conference on Robotics and Automation (ICRA), Paris, France, 31 May–31 August 2020.
41. Di Biase, L.; Perini, A.P.; Mohanty, P.; Goldenberg, A.S.; Grifoni, G.; Santangeli, P.; Santoro, F.; Sanchez, J.E.; Horton, R.; Gallinghouse, G.J.; et al. Visual, tactile, and contact force feedback: Which one is more important for catheter ablation? Results from an in vitro experimental study. *Hear. Rhythm* **2014**, *11*, 506–513. [CrossRef]

42. Wang, H.; Wang, S.; Li, J.; Zuo, S. Robotic Scanning Device for Intraoperative Thyroid Gland Endomicroscopy. *Ann. Biomed. Eng.* **2018**, *46*, 543–554. [CrossRef]
43. Ping, Z.; Zhang, T.; Gong, L.; Zhang, C.; Zuo, S. Miniature Flexible Instrument with Fibre Bragg Grating-Based Triaxial Force Sensing for Intraoperative Gastric Endomicroscopy. *Ann. Biomed. Eng.* **2021**, *49*, 2323–2336. [CrossRef]
44. Li, T.; Pan, A.; Ren, H. Reaction Force Mapping by 3-Axis Tactile Sensing with Arbitrary Angles for Tissue Hard-Inclusion Localization. *IEEE Trans. Biomed. Eng.* **2020**, *68*, 26–35. [CrossRef] [PubMed]
45. Gao, A.; Liu, N.; Zhang, H.; Wu, Z.; Yang, G.-Z. Spiral FBG sensors-based contact detection for confocal laser endomi-croscopy. *Biosens. Bioelectron.* **2020**, *170*, 112653. [CrossRef] [PubMed]
46. Zhang, W.; Ni, X.; Wang, J.; Ai, F.; Luo, Y.; Yan, Z.; Liu, D.; Sun, Q. Microstructured Optical Fiber Based Distributed Sensor for *In Vivo* Pressure Detection. *J. Light. Technol.* **2019**, *37*, 1865–1872. [CrossRef]
47. Ran, Y.; Long, J.; Xu, Z.; Yin, Y.; Hu, D.; Long, X.; Zhang, Y.; Liang, L.; Liang, H.; Guan, B.-O. Harmonic optical microfiber Bragg grating immunosensor for the accelerative test of cardiac biomarker (cTn-I). *Biosens. Bioelectron.* **2021**, *179*, 113081. [CrossRef]
48. Kumar, K.S.; Xu, Z.; Kalairaj, M.S.; Ponraj, G.; Huang, H.; Ng, C.-F.; Wu, Q.; Ren, H. Stretchable Capacitive Pressure Sensing Sleeve Deployable onto Catheter Balloons towards Continuous Intra-Abdominal Pressure Monitoring. *Biosensors* **2021**, *11*, 156. [CrossRef] [PubMed]
49. Hou, C.; Geng, J.; Sun, Y.; Chen, T.; Wang, F.; Ren, H.; Zuo, X.; Li, Y.; Liu, H.; Sun, L. A Sensorised Forcep Based on Piezoresistive Force Sensor for Robotic-assisted Minimally Invasive Surgery. In Proceedings of the 2021 IEEE 16th International Conference on Nano/Micro Engineered and Molecular Systems (NEMS), Xiamen, China, 25–29 April 2021.
50. Yu, L.; Yan, Y.; Yu, X.; Xia, Y. Design and Realization of Forceps with 3-D Force Sensing Capability for Robot-Assisted Surgical System. *IEEE Sensors J.* **2018**, *18*, 8924–8932. [CrossRef]
51. Yu, L.; Yan, Y.; Li, C.; Zhang, X. Three-dimensional nonlinear force-sensing method based on double microgrippers with E-type vertical elastomer for minimally invasive robotic surgery. *Robotica* **2018**, *36*, 865–881. [CrossRef]
52. Shi, H.; Zhang, B.; Mei, X.; Song, Q. Realization of Force Detection and Feedback Control for Slave Manipulator of Mas-ter/Slave Surgical Robot. *Sensors* **2021**, *21*, 7489. [CrossRef]
53. Zou, M.; Liao, C.; Chen, Y.; Gan, Z.; Liu, S.; Liu, D.; Liu, L.; Wang, Y. Measurement of Interfacial Adhesion Force with a 3D-Printed Fiber-Tip Microforce Sensor. *Biosensors* **2022**, *12*, 629. [CrossRef]
54. Li, B.; Gil, B.; Power, M.; Gao, A.; Treratanakulchai, S.; Anastasova, S.; Yang, G.-Z. Carbon-nanotube-coated 3D mi-crospring force sensor for medical applications. *ACS Appl. Mater. Interfaces* **2019**, *11*, 35577–35586. [CrossRef]
55. Zhan, Y.; Hong, W.; Sun, W.; Liu, J. Flexible Multi-Positional Microsensors for Cryoablation Temperature Monitoring. *IEEE Electron Device Lett.* **2019**, *40*, 1674–1677. [CrossRef]
56. Goncalves, S.B.; Palha, J.M.; Fernandes, H.C.; Souto, M.R.; Pimenta, S.; Dong, T.; Yang, Z.; Ribeiro, J.F.; Correia, J.H. LED Optrode with Integrated Temperature Sensing for Optogenetics. *Micromachines* **2018**, *9*, 473. [CrossRef] [PubMed]
57. Franz, P.; Zhu, H.; Wang, X.; Chia, R.; Hasenberg, T.; Wang, H. Tissue temperature monitoring during laser vaporization through black body radiation at wavelengths less than 1.8 um. In Proceedings of the Optical Interactions with Tissue and Cells XXXI. San Francisco, San Francisco, CA, USA, 20 February 2020.
58. Ding, H.; Lv, G.; Cai, X.; Chen, J.; Cheng, Z.; Peng, Y.; Tang, G.; Shi, Z.; Xie, Y.; Fu, X.; et al. An Optoelectronic thermometer based on microscale infrared-to-visible conversion devices. *Light. Sci. Appl.* **2022**, *11*, 130. [CrossRef] [PubMed]
59. Lv, Z.; Niu, P.; Jiang, J.; Wang, S.; Liu, Y.; Liu, T. Reflective SFT-FBG Hybrid Micro-Probe for Simultaneous Measurement of Relative Humidity and Temperature. *IEEE Photon. J.* **2021**, *14*, 6806506. [CrossRef]
60. Koh, A.; Gutbrod, S.R.; Meyers, J.D.; Lu, C.; Webb, R.C.; Shin, G.; Li, Y.; Kang, S.-K.; Huang, Y.; Efimov, I.R.; et al. Ultrathin Injectable Sensors of Temperature, Thermal Conductivity, and Heat Capacity for Cardiac Ablation Monitoring. *Adv. Health Mater.* **2015**, *5*, 373–381. [CrossRef]
61. Fujimoto, J.G. Optical coherence tomography for ultrahigh resolution in vivo imaging. *Nat. Biotechnol.* **2003**, *21*, 1361–1367. [CrossRef]
62. Hoyt, T.; Phipps, J.; Vela, D.; Wang, T.; Buja, M.; Jang, I.-K.; Milner, T.; Feldman, M. Diagnosis of Thin-capped Fibroatheromas in Intravascular Optical Coherence Tomography Images: Effects of Light Scattering. *Circulation* **2016**, *9*, e003163. [CrossRef]
63. Li, J.; Chen, Q.; Sun, J.; Zhang, J.; Ding, J.; Zuo, C. Three-dimensional tomographic microscopy technique with mul-ti-frequency combination with partially coherent illuminations. *Biomed. Opt. Express* **2018**, *9*, 2526–2542. [CrossRef]
64. Zheng, S.; Bai, Y.; Xu, Z.; Liu, P.; Ni, G. Optical Coherence Tomography for Three-Dimensional Imaging in the Biomedical Field: A Review. *Front. Phys.* **2021**, 552. [CrossRef]
65. Zhang, R.; Fan, Y.; Qi, W.; Wang, A.; Tang, X.; Gao, T. Current research and future prospects of IVOCT imaging-based detection of the vascular lumen and vulnerable plaque. *J. Biophotonics* **2022**, *15*, e202100376. [CrossRef]
66. Ono, M.; Kawashima, H.; Hara, H.; Gao, C.; Wang, R.; Kogame, N.; Takahashi, K.; Chichareon, P.; Modolo, R.; Tomaniak, M.; et al. Advances in IVUS/OCT and Future Clinical Perspective of Novel Hybrid Catheter System in Coronary Imaging. *Front. Cardiovasc. Med.* **2020**, *7*, 119. [CrossRef] [PubMed]
67. Wang, J.; Tao, K.; Zhu, W.; Jiang, J.; Liu, T. A FBG-OCT Catheter to Reconstruct Vascular Shape in Intravascular Optical Coherence Tomography. *IEEE Photon. Technol. Lett.* **2019**, *31*, 701–704. [CrossRef]

68. Kang, J.; Zhu, R.; Sun, Y.; Li, J.; Wong, K.K. Pencil-Beam Scanning Catheter for Intravascular Optical Coherence Tomography. In Proceedings of the 2020 Asia Communications and Photonics Conference (ACP) and International Conference on Information Photonics and Optical Communications (IPOC), Beijing, China, 24–27 October 2020.
69. Li, Y.; Jing, J.; Qu, Y.; Miao, Y.; Zhang, B.; Ma, T.; Yu, M.; Zhou, Q.; Chen, Z. Fully integrated optical coherence tomography, ultrasound, and indocyanine green-based fluorescence tri-modality system for intravascular imaging. *Biomed. Opt. Express* **2017**, *8*, 1036–1044. [CrossRef] [PubMed]
70. Rodriguez-Luna, D.; Molina, C.A. Vascular imaging: Ultrasound. *Handb. Clin. Neurol.* **2016**, *136*, 1055–1064.
71. Marrocco, C.J.; Jaber, R.; White, R.A.; Walot, I.; DeVirgilio, C.; Donayre, C.E.; Kopchok, G. Intravascular ultrasound. *Semin. Vasc. Surg.* **2012**, *25*, 144–152. [CrossRef]
72. Ma, X.; Cao, W. Single-Crystal High-Frequency Intravascular Ultrasound Transducer with 40 um Axial Resolution. *IEEE Trans. Ultrason. Ferroelectr. Freq. Control* **2019**, *67*, 810–816. [CrossRef]
73. Wang, L.; Lei, P.; Wen, X.; Zhang, P.; Yang, S. Tapered fiber-based intravascular photoacoustic endoscopy for high-resolution and deep-penetration imaging of lipid-rich plaque. *Opt. Express* **2019**, *27*, 12832–12840. [CrossRef]
74. Wang, X.; Peng, C.; Liu, X.; Pan, Z. Functional Assessment of Stenotic Coronary Artery in 3D Geometric Reconstruction From Fusion of Intravascular Ultrasound and X-Ray Angiography. *IEEE Access* **2018**, *6*, 53330–53341. [CrossRef]
75. Jansen, K.; Van Der Steen, A.F.; van Beusekom, H.M.; Oosterhuis, J.W.; van Soest, G. Intravascular photoacoustic im-aging of human coronary atherosclerosis. *Opt. Lett.* **2011**, *36*, 597–599. [CrossRef]
76. Su, M.; Zhang, Z.; Hong, J.; Huang, Y.; Mu, P.; Yu, Y.; Liu, R.; Liang, S.; Zheng, H.; Qiu, W. Cable-Shared Dual-Frequency Catheter for Intravascular Ultrasound. *IEEE Trans. Ultrason. Ferroelectr. Freq. Control* **2019**, *66*, 849–856. [CrossRef]
77. Zhang, Y.; Liang, Y.; Jin, L.; Guan, B. 125 μm fiber based all-optical ultrasound probes for pulse-echo imaging. *Chin. Opt. Lett.* **2019**, *17*, 070604. [CrossRef]
78. Hong, J.; Su, M.; Yu, Y.; Zhang, Z.; Liu, R.; Huang, Y.; Mu, P.; Zheng, H.; Qiu, W. A Dual-Mode Imaging Catheter for Intravascular Ultrasound Application. *IEEE Trans. Med. Imaging* **2018**, *38*, 657–663. [CrossRef] [PubMed]
79. Wang, B.; Li, X.; He, L.; Cao, T.; Zhu, C.; Wu, D. A Miniature Rotary-Linear Ultrasonic Motor for Intravascular Ultrasound (IVUS) Imaging. In Proceedings of the 2020 IEEE International Ultrasonics Symposium (IUS), Las Vegas, NV, USA, 7–11 September 2020.
80. Wilson, K.E.; Wang, T.Y.; Willmann, J.K. Acoustic and Photoacoustic Molecular Imaging of Cancer. *J. Nucl. Med.* **2013**, *54*, 1851–1854. [CrossRef] [PubMed]
81. Wang, L.V.; Yao, J. A practical guide to photoacoustic tomography in the life sciences. *Nat. Meth.* **2016**, *13*, 627–638. [CrossRef] [PubMed]
82. Wang, L.V.; Hu, S. Photoacoustic Tomography: In Vivo Imaging from Organelles to Organs. *Science* **2012**, *335*, 1458–1462. [CrossRef] [PubMed]
83. Yao, J.; Wang, L.V. Photoacoustic microscopy. *Laser Photon. Rev.* **2013**, *7*, 758–778. [CrossRef]
84. Jin, L.; Liang, Y. Fiber laser technologies for photoacoustic microscopy. *Vis. Comput. Ind. Biomed. Art* **2021**, *4*, 11. [CrossRef]
85. Du, C.; Wang, Y.; Xu, L.; Wang, X. Fiber Optic Photoacoustic-Based Steering Imaging Transducer. *IEEE Photon. Technol. Lett.* **2022**, *34*, 428–431. [CrossRef]
86. Li, G.; Ye, Z.; Liang, S.; Chen, S.L. Miniature probe for dual-modality photoacoustic microscopy and white-light micros-copy for image guidance: A prototype toward an endoscope. *J. Biophoton.* **2020**, *13*, e201960200. [CrossRef]
87. Cao, Y.; Hui, J.; Kole, A.; Wang, P.; Yu, Q.; Chen, W.; Sturek, M.; Cheng, J.-X. High-sensitivity intravascular photoacoustic imaging of lipid–laden plaque with a collinear catheter design. *Sci. Rep.* **2016**, *6*, 25236. [CrossRef]
88. Li, Y.; Lin, R.; Liu, C.; Chen, J.; Liu, H.; Zheng, R.; Gong, X.; Song, L. In vivo photoacoustic/ultrasonic dual-modality endoscopy with a miniaturized full field-of-view catheter. *J. Biophotonics* **2018**, *11*, e201800034. [CrossRef] [PubMed]
89. Shu, C.; Zheng, W.; Lin, K.; Lim, C.; Huang, Z. Label-Free Follow-Up Surveying of Post-Treatment Efficacy and Recurrence in Nasopharyngeal Carcinoma Patients with Fiberoptic Raman Endoscopy. *Anal. Chem.* **2021**, *93*, 2053–2061. [CrossRef] [PubMed]
90. Shu, C.; Zheng, W.; Wang, Z.; Yu, C.; Huang, Z. Development and characterization of a disposable submillimeter fiber optic Raman needle probe for enhancing real-time in vivo deep tissue and biofluids Raman measurements. *Opt. Lett.* **2021**, *46*, 5197. [CrossRef]
91. Shah, T.; Lippman, R.; Kohli, D.; Mutha, P.; Solomon, S.; Zfass, A. Accuracy of probe-based confocal laser endomicroscopy (pCLE) compared to random biopsies during endoscopic surveillance of Barrett's esophagus. *Endosc. Int. Open* **2018**, *06*, E414–E420. [CrossRef] [PubMed]
92. Ma, P.; Cai, L.; Lyu, B.; Yue, M. Application of probe-based confocal laser endomicroscopy in diagnosis of gastric carcinoma and precancerous lesions. *J. Zhejiang Univ. Med. Sci.* **2019**, *48*, 504–510.
93. Vennelaganti, S.; Vennalaganti, P.; Mathur, S.; Singh, S.; Jamal, M.; Kanakadandi, V.; Rai, T.; Hall, M.; Gupta, N.; Nutalapati, V.; et al. Validation of Probe-based Confocal Laser Endomicroscopy (pCLE) Criteria for Diagnosing Colon Polyp Histology. *J. Clin. Gastroenterol.* **2018**, *52*, 812–816. [CrossRef]
94. Wray, P.; Lin, L.; Hu, P.; Wang, L.V. Photoacoustic computed tomography of human extremities. *J. Biomed. Opt.* **2019**, *24*, 026003. [CrossRef]
95. Ethan Li, Y.-C.; Lee, I.-C. The Current Trends of Biosensors in Tissue Engineering. *Biosensors* **2020**, *10*, 88.
96. Singh, A.; Sharma, A.; Ahmed, A.; Sundramoorthy, A.K.; Furukawa, H.; Arya, S.; Khosla, A. Recent advances in elec-trochemical biosensors: Applications, challenges, and future scope. *Biosensors* **2021**, *11*, 336. [CrossRef]

97. Webb, B.A.; Chimenti, M.; Jacobson, M.P.; Barber, D.L. Dysregulated pH: A perfect storm for cancer progression. *Nat. Rev. Cancer* **2011**, *11*, 671–677. [CrossRef]
98. White, K.A.; Grillo-Hill, B.K.; Barber, D.L. Cancer cell behaviors mediated by dysregulated pH dynamics at a glance. *J. Cell Sci.* **2017**, *130*, 663–669. [CrossRef] [PubMed]
99. Li, Y.; Liu, X.; Xu, X.; Xin, H.; Zhang, Y.; Li, B. Red-blood-cell waveguide as a living biosensor and micromotor. *Adv. Funct. Mater.* **2019**, *29*, 1905568. [CrossRef]
100. Wang, L.; Xie, S.; Wang, Z.; Liu, F.; Yang, Y.; Tang, C.; Wu, X.; Liu, P.; Li, Y.; Saiyin, H.; et al. Functionalized helical fibre bundles of carbon nanotubes as electrochemical sensors for long-term in vivo monitoring of multiple disease biomarkers. *Nat. Biomed. Eng.* **2020**, *4*, 159–171. [CrossRef] [PubMed]
101. Guo, T.; Liu, F.; Liang, X.; Qiu, X.; Huang, Y.; Xie, C.; Xu, P.; Mao, W.; Guan, B.-O.; Albert, J. Highly sensitive detection of urinary protein variations using tilted fiber grating sensors with plasmonic nanocoatings. *Biosens. Bioelectron.* **2016**, *78*, 221–228. [CrossRef] [PubMed]
102. Ran, Y.; Xu, Z.; Chen, M.; Wang, W.; Wu, Y.; Cai, J.; Long, J.; Chen, Z.; Zhang, D.; Guan, B. Fiber-Optic Theranostics (FOT): Interstitial Fiber-Optic Needles for Cancer Sensing and Therapy. *Adv. Sci.* **2022**, *9*, 2200456. [CrossRef]

MDPI
St. Alban-Anlage 66
4052 Basel
Switzerland
www.mdpi.com

Biosensors Editorial Office
E-mail: biosensors@mdpi.com
www.mdpi.com/journal/biosensors

Disclaimer/Publisher's Note: The statements, opinions and data contained in all publications are solely those of the individual author(s) and contributor(s) and not of MDPI and/or the editor(s). MDPI and/or the editor(s) disclaim responsibility for any injury to people or property resulting from any ideas, methods, instructions or products referred to in the content.

www.ingramcontent.com/pod-product-compliance
Lightning Source LLC
LaVergne TN
LVHW070541100526
838202LV00012B/345